THE MANUFACTURE OF ORGANIC COLORANTS AND INTERMEDIATES

GERALD BOOTH

SOCIETY OF DYERS AND COLOURISTS

PO BOX 244, PERKIN HOUSE, 82 GRATTAN ROAD, BRADFORD BD1 2JB

1988

Typeset by Wessex Typesetters
(Division of The Eastern Press Ltd)
Frome, Somerset

Printed in Great Britain by
The Eastern Press Ltd
London and Reading

ISBN 0 901956 47 3

Acknowledgement

The author is grateful for cooperation received from Bayer, Ciba–Geigy, Hoechst, Sandoz and Yorkshire Chemicals through the provision of published data and photographs. In addition, many equipment suppliers provided material for inclusion and individual acknowledgements are made with the relevant figures. Finally thanks are due to many colleagues in Organics Division of ICI who have been of enormous help in providing data and comments along with reading the initial draft chapters.

THE AUTHOR:

Gerald Booth, formerly research manager, ICI plc, Organics Division, Blackley, Manchester
Present address: Thorpe House, Uppermill, Oldham OL3 6DP

Previous titles in this series published by the Society of Dyers and Colourists:

1. A practical introduction to yarn dyeing, by James Park
2. A practical introduction to fibre and tow coloration, by Geoffrey Clarke
3. Dyehouse laboratory practice, by T M Thompson
4. An introduction to management in the dyeing industry, by James Park
5. A practical introduction to the dyeing and finishing of wool fabrics, by Ian Bearpark, F William Marriott and James Park

Contents

CHAPTER 1

Introduction

1.1 THE TECHNOLOGY

The technology of dyes and intermediates, in total, is a combination of the following components:

(a) Product
(b) Application
(c) Process technology
(d) Plant technology,

and each component may be subdivided so that inter-company comparisons may be made. Manufacturing technology may conveniently be considered as a combination of process and plant technologies.

Process technology involves chemists and chemical engineers and will be discussed with a bias towards unit processes (i.e. sulphonation, nitration, reduction, etc.). With plant technology the chemical engineer links with the mechanical engineer to design and construct manufacturing units that are then operated by technical plant managers of various scientific disciplines. In a parallel way to process technology, it is usual to consider plant technology from a unit operations (i.e. mixing, filtration, drying, etc.) viewpoint.

Previous publications dealing with the links between these and supporting technologies appeared over 30 years ago [1, 2] and a major problem in updating this information is the scarcity of material published in the meantime. Detailed technological know-how is carefully guarded, since once knowledge of the basic chemistry is available, and the products lie outside any patents, then manufacturers need to safeguard the confidentiality of these products and their processes. Patenting of process and operational technology is comparatively rare in the dye-manufacturing industry because of the risk of disclosure and the difficulty of enforcement. The best detailed source of process information still remains the BIOS and FIAT reports describing German manufacture up to 1945.

The purpose of this monograph is to describe the manufacture of organic colouring matters (dyes, pigments and fluorescent brightening agents (FBAs)) and of the organic intermediates used to make them. These intermediates include all chemicals from which dyes are made, only excluding raw materials such as hydrocarbons and bulk chemicals already required by the rest of the chemical industry (e.g. benzene, toluene, chlorobenzene, naphthalene, anthracene, phenols, alcohols and other solvents). Although not strictly connected, this total package of manufacturing components will be called the dyestuffs industry.

Inorganic products such as iron oxide and chrome pigments are outside the scope of this monograph and are not discussed here. A good summary of their manufacture is available in the *Pigment handbook* [3].

Significant quantities of textile auxiliaries are also manufactured within the dyestuffs industry but these are usually considered as a separate business and also are not within the scope of this review. For further reading on the manufacture of all types of surface-active agents the book by Schwarz and Perry [4] is recommended, while the book by Schönfeldt [5] is relevant to the manufacture of ethylene oxide adducts.

It should also be remembered that dyestuffs technology has been instrumental in the launching of whole new industries. For example, many pharmaceuticals and agrochemicals have arisen from its organic chemistry product base, and many new synthetic materials from its manufacturing and process technology base.

1.2 THE CONSUMER

Everybody is a consumer of coloured products. Throughout the world consumption of organic colouring matters is around 200 g per person per annum, although in developed countries this figure is much higher, e.g. in West Germany the average is 650 g per annum. Of course products from clothing to cars are involved, and an interesting statistic is that 1 tonne of dye colours 41700 suits and 1 tonne of pigment colours 16 600 VW Beetles [6].

Many different industries use dyes and pigments; the chief ones are textiles, leather and paper, with pigments being used mainly for paints (and other surface coatings including textile printing), inks, plastics and rubber. Textiles dominate in consuming about 75% of the world output of colorants, representing a total market value of over £2 500 million per annum. Current estimates suggest that the cost of dyes and chemicals represents about 20% of the total cost of textile dyeing and finishing, equivalent to about 4% of the cost of the final garment. Specialised applications include colorants for food, photography, fur, hair and electronics; manufacture of products for these purposes is sometimes undertaken by the mainstream dyestuffs industry, but may alternatively be carried out in the consuming industry.

Different dye classes are used for different types of fibre, so the nature of the dyestuffs industry is inextricably linked with that of the textile industry. The relative importance of the major fibre types is shown in Table 1.1. The total production of synthetic-polymer fibres has grown rapidly in recent decades and now approaches that of the natural fibres (see Figure 1.1).

The well-being of the dye-manufacturing industry is clearly dependent on the fortunes of the fibre manufacturers, so Figure 1.1 can also be regarded as broadly reflecting the growth on demand for the various dye classes required (note logarithmic scale). Figure 1.1 also shows the boom period in the early 1970s, which at that time gave rise to over-optimistic forecasts of the production capacity required in both the fibre- and dye-manufacturing industries.

TABLE 1.1

World production of fibres in 1980

Fibre	*Production (tonnes × 10^3)*
Cotton	14 100
Flax	500
Wool	1 600
Silk	50
Viscose/acetate	3 600
Polyester	5 200
Polyamide	3 200
Acrylic	1 700

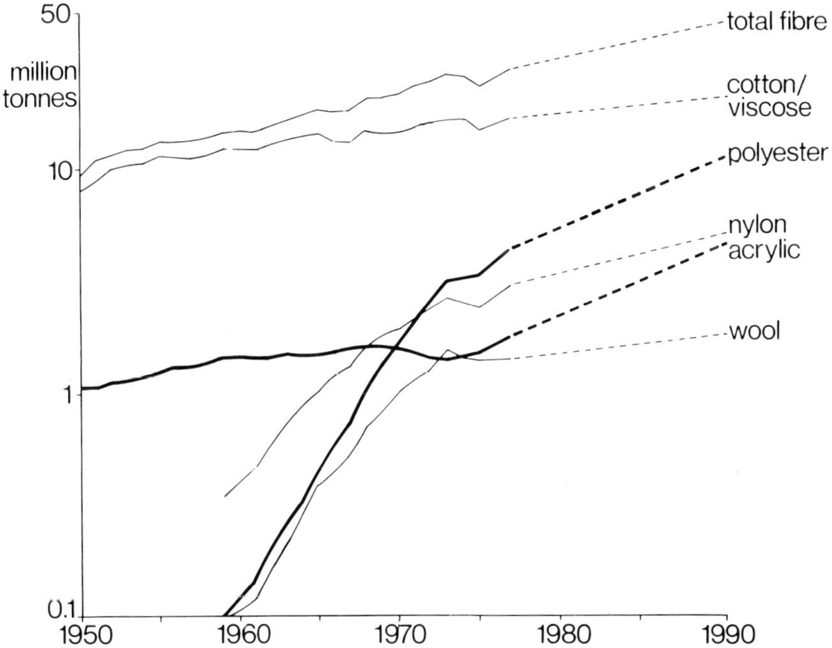

Figure 1.1 – Estimated world fibre production

Accurate forecasting of future fibre production will, of course, always be important to the dye manufacturer, but there are other variables in the econometric equation. A recent detailed analysis of the US market [7] has shown a considerable reduction in the amount of dyes consumed as a proportion of fibre consumption between 1965 and 1981. When dye consumption decreases in relation to fibre consumption the four main factors involved are:

(a) Fashion trends favouring pale colours and white at the expense of deep colours and blacks
(b) Use of improved application technology to give better yields on the substrate, e.g. a trend from batchwise to continuous processing
(c) Optimisation of formulations and physical forms
(d) Use of stronger dyes with either increased inherent colour value or simply increased brand strength (the latter is especially relevant to actual weight comparisons of disperse dyes).

The relative contributions of each of the above factors cannot be established but there is no doubt that research and development (R & D) efforts over the last decade have increasingly concentrated on the last three.

Statistics that quote actual weights of dyes manufactured should be viewed with caution. Dyestuffs are subject to variations in content of active agent; they rarely represent 100% dye content. The financial worth of the dye business is what matters to the manufacturer, but comparisons in value terms can also mislead during inflationary periods.

1.3 THE READER

This monograph is aimed principally at those working or interested in the coloration industries but who may not be directly associated with dye manufacture. The chemistry of dyes and intermediates is not discussed in detail as this is well covered in books such as those edited by Venkataraman [8], whereas their manufacture is not. Against a background of a constantly changing manufacturing scene, discussion of subjects such as the structure of the industry reflects data published up to the end of 1984.

REFERENCES

1. Fierz-David and Blangey, 'Fundamental processes in dye chemistry' (New York: Interscience, 1949).
2. Frank, 'The manufacture of intermediates and dyes' (Constable, 1950).
3. 'Pigment handbook', Vol. 1, Ed. Patten (New York: John Wiley, 1973).
4. Schwartz and Perry, 'Surface active agents', Vols. 1 and 2 (New York: Interscience, 1948, 1958).
5. Schönfeldt, 'Surface active ethylene oxide adducts' (Oxford: Pergamon, 1969).
6. Boresch, Defazet, **33** (6/7) (1979) 25.
7. Kuehni, Text Chem. Colorist, **15** (2) (1983) 20.
8. Venkataraman, 'The chemistry of synthetic dyes', 8 vols. (New York: Academic Press, 1952–78).

CHAPTER 2

Basic structure and economics of the dyestuffs industry

2.1 WORLD PRODUCTION

By the early part of the 20th century about 85% of world dyestuff requirements was manufactured in Germany, with other European countries (Switzerland, UK and France) accounting for a further 10%. Seventy years and two world wars have seen dramatic changes in this pattern. Table 2.1 shows that, in weight terms, the Western European share of world production had moved to 50% by 1938 and 40% by 1974. However, since a large part of US manufacture and some of 'others' is based on Western European subsidiaries, then their overall share remains at over 50%.

The figures in Table 2.1, showing progress through the last five decades, are based on published estimates [1,2]. Since 1974 many of the national figures have not been published, but those that are available indicate that 1974 was the peak production year of the 1970s, with 1975 being the nadir. World recession caused a 20% slump in production which has now been recovered. Consequently the figures of Table 2.1 are still applicable to the present day within their limits of accuracy. More recent independent reviews by O'Sullivan [3] and Schundehutte [4] confirm a world output of 750 000 to 800 000 tonnes per annum.

TABLE 2.1

World dyestuff production estimates

	Production (tonnes × 10^3)					
Year	*W Europe*	*USA*	*USSR E Europe China*	*Japan*	*Others*	*Total*
1938	110	37	35		28	210
1948		110				
1958	112	80	127		27	346
1966	191	130		49		
1974	300	138	200+	68	44	750+

With the present state of the world economy the growth rate for the industry is likely to be as low as 2–3% through the 1980s. However, even this low figure represents something like 20 000 tonnes per annum additional capacity requirement.

It is important to bear the distribution of production in mind when more detailed national analyses are discussed. Clearly most countries are net importers of dyes, but even major manufacturing countries have significant imports.

2.1.1 Western Europe

A country by country analysis of European dyestuff manufacture is given in Table 2.2 together with respective values. A clear example of the need to consider price as well as volume is obtained by comparing production in Switzerland and France. Each country manufactured roughly the same tonnage but average 1974 values were respectively US $12.3 per kg and $4.75 per kg against an overall European selling price average of $6.3 per kg. The higher than average price for Swiss dyestuffs reflects a combination of higher-strength products and an emphasis on more expensive speciality products.

TABLE 2.2

West European dyestuff production estimates (1974)

Country	*Production (tonnes × 10^3)*	*Value (US $ × 10^6)*
West Germany	148	976
UK	54	268
Switzerland	30	370
France	31	147
Italy	15	62
Spain	15	42
Belgium	6	23
Rest	1	4
Total	300	1 892

Mention of individual companies in the following sections is limited to major manufacturers, arbitrarily defined as those belonging to the Ecological and Toxicological Association of the Dyestuffs Manufacturing Industry (ETAD) [5]. It is acknowledged that there are many more smaller dye manufacturers but these typically have turnovers of less than £10 million per annum.

UK

Over 90% of UK manufacture is undertaken by ICI plc Organics Division, Yorkshire Chemicals plc and subsidiary operations of the German and Swiss majors. Such subsidiary production is typical of the worldwide scene in which the multinationals use local plants to increase their options for complete or part manufacture of individual products.

The optimism of thc early 1970s inspired considerable UK investment in both intermediates and dyestuff (mainly azo) capacity. The overcapacity resulting from the subsequent recession seems likely to persist through the 1980s.

In 1982 the largest UK producer ICI acquired the major French manufacturer, PCUK (Colours Division), forming ICI Francolor, which is now integrated with the ICI Organics Division.

West Germany

Divergent post-war developments within the three German majors Bayer, BASF and Hoechst (including Cassella which is 75% owned by Hoechst) have led to contrasting profiles. In common with ICI, development of their dyes businesses, representing less than 5% of a total chemicals business, must be seen in the wider context of both heavy chemicals (including plastics and fibres) and fine/speciality chemicals (including dyes).

Bayer's strength in intermediates has manifested itself in a DM 500 million joint investment with Ciba–Geigy for new capacity (see Chapter 4). This Schelde-Chemie plant at Brunsbuttel is by far the largest single investment in new dye/intermediate capacity that the Western world has seen since its main manufacturing bases were established 50 or 60 years ago. Establishment of a major new works on the Elbe estuary will significantly lessen the effluent load on the Rhine, close to which most of the Swiss and the German industries have been based.

Switzerland

Unlike the Germans and the British, the Swiss have confined themselves to their original fine chemicals base, so that dyes represent 20–30% of their chemical industry. This must, however, be continually decreasing in the face of more rapidly expanding pharmaceuticals and agrochemical sectors. They export 80% of their dye production, compared with 65% for UK and 50% for Germany.

Since 1970 Ciba-Geigy (formed by the merger of CIBA SA and J R Geigy SA) and to some extent the other Swiss major Sandoz have been going through a rationalisation phase. The Swiss are dependent upon imports for many of their intermediates and the Ciba-Geigy investment in Schelde-Chemic, referred to earlier, reflects the need to ensure their long-term supply. Sandoz announced a similar joint venture with BASF in 1974 but this was subsequently abandoned.

Italy

ACNA (a Montedison subsidiary) was until recently the major manufacturer, but in spite of investments in new environmentally improved plants, its lack of profitability had been crippling ever since the onset of recession in the textile industry. In 1983 the textile dyes operation was closed down and the pigments manufacturing facilities (3 000 tonnes per annum) sold to Inmont Corporation of the USA. More so than other European manufacturers, ACNA has sold or licensed much of its intermediate and dye technology to Eastern European countries.

2.1.2 USA

The US scene has remained much more fragmented than that of European countries. Manufacture is shared between at least ten manufacturers rather than three or four for each European country. All the European majors have a significant manufacturing presence in the USA, gradually replacing exports from Europe.

The demise of the long-standing DuPont dye business in 1979 (shortly after a major investment in a new azo dye plant in Puerto Rico in 1974), the withdrawal by American Cyanamid in 1980, the liquidation of the American Color and Chemical business in 1982, and the takeover of the Sodyeco Division of the Martin-Marietta Corporation by Sandoz in 1983 have left Atlantic Chemical and Crompton & Knowles as virtually the last locally owned manufacturers with wide product ranges. Atlantic formed a joint marketing venture with Mitsubishi Chemical Industries in 1982 to provide a significant Japanese entry into the American market.

Other American firms should be mentioned for their more specialised activities:

(a) Eastman Chemical Products, a significant producer of disperse dyes, with a restricted approach stemming from its unique integration with a dominant photographic colours manufacturer (Eastman Kodak)
(b) Buffalo Color, which in 1981 withdrew from the dyestuffs business except for production and sale of indigo
(c) Sun Chemical, pigment manufacturer
(d) DuPont, which retained its intermediates business, but announced that it was ceasing pigment manufacture from 1984.

2.1.3 Japan

A historically fragmented Japanese industry, with rapid growth to meet internal demand, changed in 1979 with an agreement by its five major manufacturers (Mitsubishi, Sumitomo, Nippon Kayaku, Mitsui Toatsu and Hodogaya) to the formation of a cartel to coordinate research as well as manufacture. Successful operation of this cartel could have profound effects on the international situation by the end of the century.

2.1.4 USSR, Eastern Europe and China

Comparatively little is known of manufacture in these countries, although their aim must be to satisfy all their own requirements. The signs are that they need to import technology to hasten the process towards self-sufficiency. They are currently significant importers of dyes, although Czechoslovakia in particular, more recently joined by China, does export a limited number of intermediates.

It is noteworthy that these countries have so far been the only ones to announce large-scale investment in new dye-manufacturing capacity for the 1980s. These are a 4 000 tonnes per annum disperse dye plant and a 2 000 tonnes per annum leather dye plant in the USSR, both using Montedison/ACNA technology. The Italian group is also reported to be providing know-how for a dyestuff finishing plant to be built in north-east China.

2.1.5 Rest of the World

Manufacture in India, South America and Mexico is mainly through subsidiaries, or partly owned companies associated with the European majors. Within Far Eastern countries, other than Japan, a large number of small manufacturing facilities are evolving, especially in Taiwan and South Korea.

2.2 PRODUCT RANGES

Details of the production of the various colorant classes is available only for the USA [2] and Japan [6]. These are summarised in Table 2.3 not just to

TABLE 2.3

US and Japanese production of colorants in 1980 (tonnes)

Class	*Japan*	*USA*
Acid	2 300	11 600
Basic	5 000	6 600
Direct	3 200	14 000
Disperse	13 600	21 000
Reactive	4 700	2 600
Fluorescent brightener	7 500	17 000
Mordant	1 500	200
Solvent	3 000	4 800
Vat	2 400	18 000
Other (azoic, sulphur, food, etc.)	8 200	15 200
Total dyes	51 400	111 000
Organic pigments	21 000	31 000
Total dyes and pigments	72 400	142 000

compare the two countries, but to permit a very rough estimate of the world manufacturing split by multiplying the sum of the output for these two countries by four (cf. Table 2.1).

Estimates of the consumption of dyes for textiles (about 60% of the total dyestuff market world-wide by volume) are given in Figure 2.1, together with figures for fibre consumption. This shows that the consumption of each fibre type is approximately in agreement with the consumption of the principal dye class(es) used for it [7].

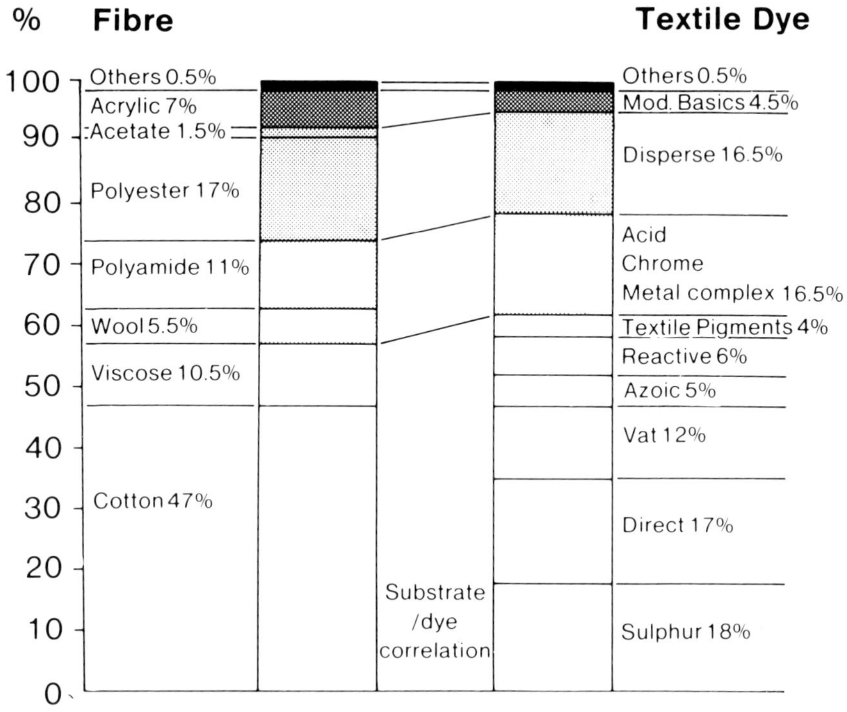

Figure 2.1 – Comparison of fibre and textile dye consumption (1979 estimates)

Table 2.3 shows clearly the differing patterns of manufacture. The very different ratios for vat and reactive dyes in the two countries, for example, reflect the wide range of market needs that are possible. For the US, costs of colorants average US $7.7 per kg, ranging from low-cost FBAs at $4.0 per kg to reactive dyes at $14.3 per kg.

2.3 COSTS

A recently published analysis of manufacturing costs compares two alternative approaches to dye manufacture [8]. In Table 2.4 case A represents a high added-value situation, starting from basic chemical building blocks such as benzene and naphthalene. Case B represents the alternative of starting from late-stage intermediates, with activity confined to the final stage of colorant manufacture. Many, but not all, major manufacturers tend towards case A, and small converters to case B, but various equilibria are reached between A and B for many reasons other than scale of operation.

TABLE 2.4

Dyestuff cost profile

	Cost (£ per kg)	
	Case A	*Case B*
Raw materials	0.7	1.7
Energy costs of manufacture	0.2	0.1
Manufacturing expenses (salaries/wages)	0.8	0.4
Administration costs (R & D, selling expenses, etc.)	0.8	0.55
Servicing fixed and working capital	0.5	0.25
Total	3.0	3.0

Apart from showing the range of cost structures possible, these hypothetical figures can be used to derive the following conclusions.

(a) Raw materials are a significant proportion of cost and are outside the control of the manufacturer (other than his efficiency of conversion), with inflation (including world oil price rises) having an immediate impact on their cost.

(b) Fuel (i.e. direct energy as distinct from the energy already paid for in the price of the raw materials) is a comparatively small proportion of the high added value, but also is subject to the same cost pressures as raw materials. There is always a need to reduce fuel consumption but savings will be small in relation to the total cost.

(c) The cost elements under the manufacturer's control, most simply stated as manpower costs and capital costs, are clearly a function of his raw material starting-point. In case A manpower costs are about 50% of the total, whereas in case B they are about 30%. Obviously, if all intermediates are purchased, much less plant is required and thus there is less capital to be serviced, but in either case the efficiency of capital utilisation (expressed in, for example, kg produced per £ of capital) is critical. It is worth noting that the working capital required by these modes of operation can also differ widely.

(d) The arguments for and against cases A and B are so important as to lead manufacturers to assess each product range, and maybe even individual products, by some such comparison. With an increasing number of basic intermediates moving into the category of being non-specific to dyes, intermediate costs and policies on manufacture become more critical (cf. Chapter 4).
(e) The significance of R & D can be derived from this analysis. All the European major manufacturers, unlike, for example, the Americans, have maintained a high level of technical expenditure on R & D, generally estimated as 4–5% of turnover if the high level of technical service required to support an international business is included [4]

The difference, or margin, between the (hypothetical £3 per kg) cost and the dye selling price (also referred to as the unit value) represents the profit (or loss) to the manufacturer. Most of the industry has suffered at least temporary losses since the 1974 boom. A survey of the American industry [9] highlights the four following reasons why otherwise successful chemical companies have withdrawn from the dyes sector.
(a) There has been considerable overcapacity in the industry, with a subsequent inability to increase prices sufficiently to recover rising raw material and labour costs, and thereby to generate profit.
(b) The complexity of dyes (some taking as long as four months to produce) means that dye inventories turn over only two or three times a year. This means that dye production is unusually capital intensive. Growth (or change) is expensive particularly if money has to be borrowed to support that growth.
(c) The associated highly specialised technical service function adds further to the operating costs.
(d) Government regulations have greatly increased operating expenses and forced dye companies to spend large amounts on effluent treatment and air pollution abatement.

REFERENCES

1. OECD chemical industry statistics.
2. US International Trade Commission statistics on production and sale of dyes and pigments.
3. O'Sullivan, Chem. and Eng. News (26 Feb 1979) 16.
4. Schundehutte, Defazet, **3** (1979) 202.
5. Clarke, Text. Chem. Colorist, **15** (3) (1983) 42.
6. Japan Chemical Annual (1981).
7. Aeberhard, Textilveredlung, **16** (1981) 442.
8. Gow and Booth, J.S.D.C., **97** (1981) 154.
9. Sullivan, Amer. Dysestuff Rep., (1981) 44.

CHAPTER 3

Raw materials and primary intermediates

3.1 DEFINITION

The term raw materials is frequently employed to describe the starting materials, such as benzene, chlorobenzene and naphthalene, used in many chemically based industries. Primary intermediates (henceforth referred to as primaries) are also discussed in this chapter since, in common with raw materials, they are also used in industries other than dyestuffs. Primaries are characterised by one or more of the following descriptions, which associate them with raw materials rather than with intermediates:

(a) Manufactured in a dedicated plant, i.e. one devoted to a single product or at most two or three closely related products
(b) At least 1 000 tonnes per annum capacity from a single plant and may be up to 100 000 tonnes per annum (e.g. aniline)
(c) Manufacturing process and/or operation is continuous or semi-continuous (i.e. at least one stage is in a continuous, as distinct from batch, mode)
(d) Established usage in basic industries such as rubber, polymers or agrochemicals in addition to dyestuffs.

3.2 RAW MATERIALS

These may be conveniently discussed under the three broad headings: inorganics, coal-based and oil-based materials.

3.2.1 Inorganic raw materials

These include acids (sulphuric, nitric, hydrochloric and phosphoric), bases (caustic soda, caustic potash, soda ash, sodium bicarbonate, ammonia and lime), salts (sodium chloride, sodium nitrite and sodium sulphide) and other substances such as chlorine, bromine, phosphorus chlorides and sulphur chlorides. The important points to bring out are that:

(a) There is a significant usage of at least one inorganic material in all processes and the overall tonnage used by, and therefore the cost to, the industry is high
(b) Handling and storage are especially important in view of corrosiveness as well as bulk
(c) Even though inorganics are cheap they are significant contribution to the raw material cost of final products

(d) In periods of inflating energy costs (when naphtha prices do not necessarily rise to the same extent) there is a disproportionate increase in the cost of inorganics since the energy component in their production is high, e.g. the cost of caustic soda is directly related to the cost of electricity.

3.2.2 Coal-based raw materials

The birth of the synthetic organic dyestuffs industry in 1856 was followed by a total dependence on coal tar as a source of aromatics for virtually the first 100 years. Since the 1950s petroleum-derived naphtha has taken over as the major source of monocyclic aromatics; coal tar, however, continues to be the major source of polycyclic aromatics. Production of coal (town) gas, the alternative source of coal tar, came to an end in most developed countries in the 1960s. All coal tar, about 4% recovery by weight, now derives from the coke consumed in blast furnaces and is therefore wholly dependent upon the location, size and possible future technology of the iron and steel industry.

Details of the complex distillation, crystallisation and extraction processes associated with coal tar are confined to the summary in Figure 3.1 [1]. A guide to the quantities of the relevant constituents is given in Table 3.1 together with their major uses in the dyestuffs industry.

The aim of tar refining is to extract as much as possible of its individual constituents whilst making the remaining oils and residues suitable for specific applications. Distillation is the first step, followed by crystallisation or hydrofining.

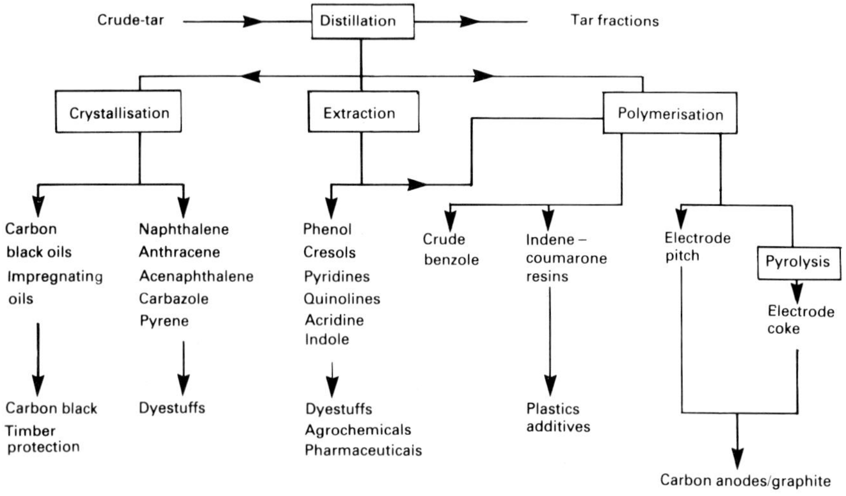

Figure 3.1 – Processing of coal tar

TABLE 3.1

Constituents of coal tar and their uses in dye manufacture

Compound	*Formula*	*Amount (%)*	*Use in dye manufacture*
Naphthalene	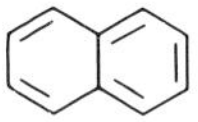	10	β-Naphthol Letter acids* Phthalic anhydride
Anthracene	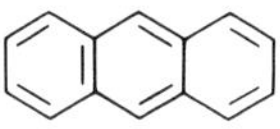	1.8	Anthraquinone
Pyrene	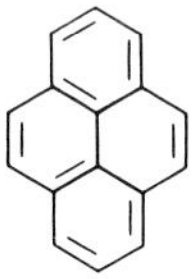	2	Naphthalene-1,4,5,8-tetracarboxylic acid FBAs
Acenaphthene	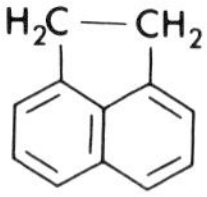	0.5	Naphthalic anhydride
Carbazole	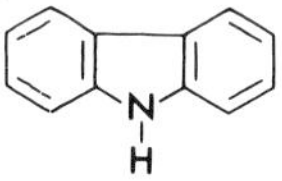	1.5	2-Hydroxycarbazole
Cresols	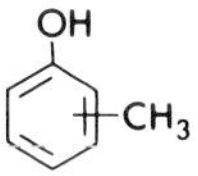	0.5	Azo dyes

* Letter acids is the collective term used to describe a large number of important substituted naphthalene sulphonic acids, e.g. H-, J-, γ-, C-, G- and R-acids (see Chapter 4 and Glossary)

In the case of naphthalene, sulphur compounds (e.g. thionaphthene) are removed in the crystallisation process and a high-quality product (m.p. 80.1°C) is obtained. World-wide production of naphthalene is estimated at 1 million tonnes per annum. Most of it is oxidised to phthalic anhydride with approximately 15% converted to naphthalene intermediates (the usual terminology to cover the whole range of derivatives from β-naphthol through to the various letter acids).

Anthracene is the second most commercially important component but it is technically more difficult to extract and purify since the associated compounds phenanthrene and carbazole have similar boiling points. Compared with a 90% recovery of naphthalene, a 40–50% recovery of anthracene appears to be the best that can economically be achieved. This accounts for the recent drive to compare alternative routes for making anthraquinone (see Section 3.3.4).

The amount of pyridine present in coal tar (0.02%) has for a long time been insufficient to meet demand, and so pyridine has to be synthesised from alternative sources. Cresols, on the other hand, are supplied by coal tar distillation and via synthetic routes.

The potential of coal tar as a source of condensed aromatic chemicals is not fully exploited; uses for phenanthrene (5%), fluoranthrene (3.3%) and fluorene (2%) are of insufficient importance to justify their relatively costly extraction [2].

3.2.3 Oil-based raw materials

Apart from the polycyclic products from coal tar mentioned above, the world has become increasingly dependent on petroleum and natural gas for organic chemical feedstock. Production of benzene, toluene and xylene from coal tar has become uneconomic over the last 20 years, although the re-emergence of 'king coal' is predicted for the 21st century when oil supplies are expected to dwindle and the cost of producing chemicals from coal (including new processes based on synthesis gas) will gradually become more competitive [3].

TABLE 3.2

Petrochemicals and derivatives

Hydrocarbon	*Derivatives*
Aliphatic	
Ethane	Ethylene, ethyl chloride
Ethylene	Ethylene oxide, ethylene glycol, vinyl chloride, ethanol
Propane	Propylene, acetic acid
Propylene	Acrylic acid, acrylonitrile, propylene oxide, isopropanol, acetone, glycerol, cumene
Butadiene	
Cyclic	
Cyclohexane	Adipic acid, caprolactam
Benzene	Ethylbenzene, cumene, cyclohexane, nitrobenzene, chlorobenzene, maleic anhydride, alkylbenzenes
Toluene	Benzene, benzoic acid, (di)nitrotoluenes
p-Xylene	Terephthalic acid
o-Xylene	Phthalic anhydride
Ethylbenzene	Styrene
Cumene	Phenol, α-methylstyrene, acetophenone, acetone

The naphtha fraction from petroleum refining is currently subjected to cracking and catalytic reforming processes resulting in aromatisation to levels of up to 50%. The petrochemicals industry has an advantage over the coal tar industry as a supplier of simple aromatics because of its flexibility in balancing its fractionation and reforming processes to meet changing patterns of demand.

Table 3.2 shows the key petrochemical hydrocarbons and some of their derivatives used as raw materials, without attempting to cover the associated manufacturing technology.

There can be little doubt that ethylene, propylene, benzene and toluene emerge as the most important petrochemicals. Together with naphthalene and anthracene they represent the organic feedstock for over 90% of the dyestuffs industry. Of the few organic raw materials outside this group there should be mentioned aliphatic amines (derived from the corresponding alcohol) and cyanuric chloride, widely used in FBAs and reactive dyes, and manufactured by trimerisation of cyanogen chloride (ClCN).

3.3 PRIMARIES

To the description given in Section 3.1, it should now be added, to justify a more detailed approach, that primaries were originally manufactured within the dyestuffs industry. All the significant primaries, about 30 different

TABLE 3.3

US and Japanese production of primaries in 1980 (tonnes)

Product	*USA*	*Japan*
Nitrobenzene	278 012	50 961
Aniline	299 737	84 661
Dimethylaniline	*(3)	1 504
Phenol	1 167 050	212 173
Salicyclic acid	17 749	2 338
o-Chloronitrobenzene	*(2)	10 463
p-Chloronitrobenzene	*(2)	20 262
m-Phenylenediamine	*(1)	
p-Phenylenediamine	*(2)	384
p-Nitroaniline	6 558	
o-Nitrotoluene	*(2)	4 534
p-Nitrotoluene	*(2)	2 872
o-Toluidine	*(1)	1 832
p-Toluidine	*(1)	1 399
Benzoic acid	33 370	
Phthalic anhydride	371 930	219 943
β-Naphthol	*(1)	4 485

* Manufactured (number of producers in parentheses) but quantities not disclosed

products, are derived from benzene, toluene or naphthalene. The syntheses will be shown schematically and then described individually, since all the primaries together with anthraquinone are the key building blocks of the industry.

Actual production figures for primaries are not readily available and in any event the amounts used within the dyestuffs industry are variable. Nevertheless the available figures from the USA [4] and Japan [5] are given in Table 3.3 so that some rough comparison may be made with those countries' production of dyes, as given in Table 2.3. The following conclusions can be drawn from Table 3.3:

(a) Not all aniline is derived from nitrobenzene
(b) The *para*:*ortho* ratio in the nitration of chlorobenzene is approximately 2:1
(c) The *para*:*ortho* ratio in the nitration of toluene is approximately 2:3
(d) Approximately half the quantities of the nitrotoluenes are reduced to toluidines.

3.3.1 Benzene derivatives

The primaries made from benzene or its derivatives are shown schematically in Figure 3.2 to demonstrate the principal manufacturing routes. Where names are given (sometimes accompanied by abbreviations in established usage, see Glossary), this indicates that the products will be considered in more detail.

Aniline (I: Figure 3.2)
Annual world production of aniline is many hundreds of thousands of tonnes. Most of this is used for the manufacture of polyurethane products, but about 5% of annual production is consumed in dye making.

The conventional highly efficient process (see Figure 3.3) for making aniline is through nitration of benzene to nitrobenzene, followed by reduction to aniline. Both the nitration and the reduction processes have many variants and the technologies are applicable to associated compounds, e.g. toluidines.

Nitration of benzene is carried out on a large scale, mainly to meet the requirement for aniline, thus allowing the capital investment involved and the economics of operation to be optimised. The flow diagram shown in Figure 3.4 is self-explanatory and shows the importance of recycling reactants such as spent sulphuric acid. Each of the boxes in the diagram represents a stage in the process with its own developed technology. The most recent refinement of this process is adiabatical operation, utilising the heat of reaction for the reconcentration of the sulphuric acid [6].

Catalytic reduction is widely used but has not completely supplanted iron reduction processes, even with continuous operation. Sometimes the necessary cheap hydrogen is not available, but one advantage of the latter

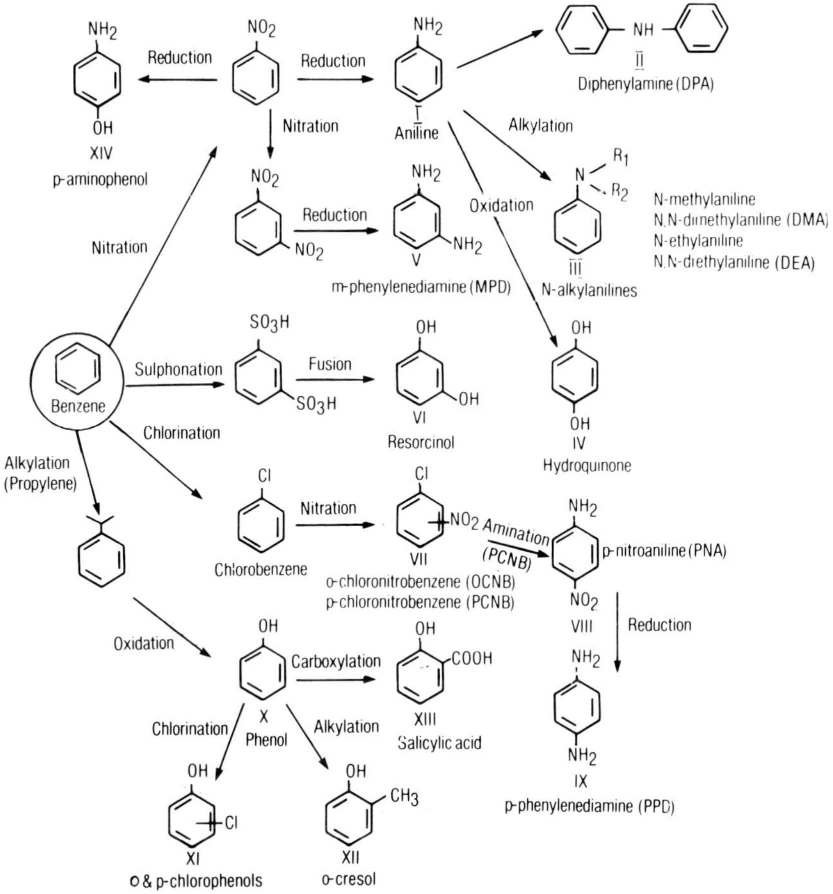

Figure 3.2 – Manufacture of primaries from benzene and chlorobenzene: industrial routes

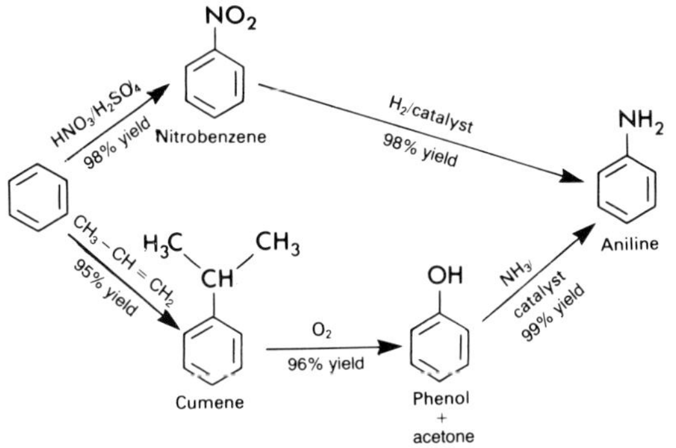

Figure 3.3 – Alternative manufacturing processes for aniline

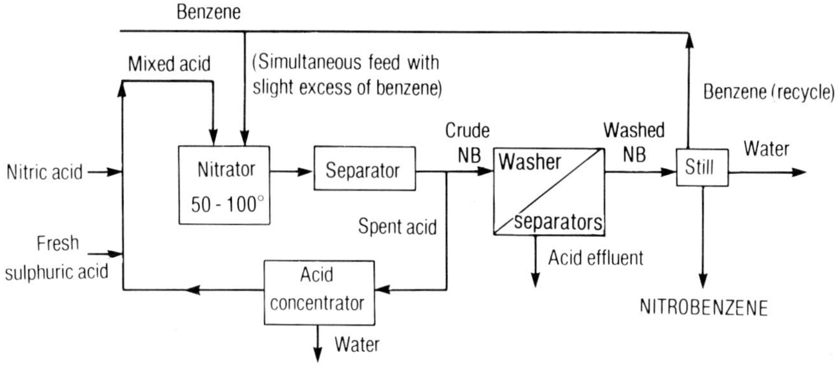

Figure 3.4 – Manufacture of nitrobenzene by the continuous process

process is the possibility of simultaneously producing pigment-grade iron oxide, which alters the economic comparison. Earlier hydrogenation processes were based on vapour-phase reaction using a silica-supported copper catalyst. The alternative ICI process [7] uses the less energy-intensive approach of liquid-phase reaction with a supported nickel catalyst.

The main alternative (Halcon) process, based on ammonolysis of phenol, has been compared in detail with the conventional process, but although proven technically it still requires continuing economic appraisal [8].

Diphenylamine (II: Figure 3.2)

The old method for making diphenylamine involving the reaction of aniline with its hydrochloride at 230°C has been superseded by a catalytic process. Aniline vapour is passed over activated alumina at 400–500°C to give diphenylamine plus ammonia in high yields. A product of high purity is obtained after passage through a distillation stream. Much more diphenylamine is used in the manufacture of rubber chemicals than of dyestuffs.

N-*Alkylanilines* (III: Figure 3.2)

Modern plants are available to operate the Thoma process [9] for the continuous alkylation of aniline with methanol or ethanol and a catalyst to form mixtures of the corresponding monoalkyl- and dialkyl-anilines. Proportions of secondary and tertiary arylamine can be partly controlled by reactant ratios, and the products are separated by fractional distillation.

Hydroquinone (IV: Figure 3.2)

Oxidation of aniline remains the main process of hydroquinone manufacture, although hydroxylation of phenol, giving a mixture of catechol and hydroquinone, is a more recently developed alternative. Excess manganese dioxide converts aniline to benzoquinone which can be separated by steam distillation and reduced catalytically using hydrogen, or by sulphur dioxide or iron. A greater proportion of world production of hydroquinone, amounting to 40 000 tonnes per annum, is used in photographic and rubber chemicals.

m-*Phenylenediamine* (V: Figure 3.2)
m-Dinitrobenzene (DNB) is first manufactured by further nitration of nitrobenzene. The second nitration is performed on isolated nitrobenzene rather than attempting a through process since significantly different conditions (100°C vs 50°C) in more concentrated sulphuric acid are required. The yield of DNB is 85% after purification, by sulphite or caustic treatment.

For most manufacturers subsequent iron reduction of DNB has been replaced by catalytic reduction. The costs of hydrogen/catalyst processes are especially favourable for diamines since, unlike iron reduction where the quantity of iron determines the batch size, two nitro groups in one molecule are only marginally more costly to reduce than one.

Resorcinol (VI: Figure 3.2)
A well established, and somewhat inelegant, process is still used for the manufacture of resorcinol. Benzene is monosulphonated with 100% sulphuric acid (monohydrate) at 80°C and the product converted to benzene-1,3-disulphonic acid by adding 30% oleum at 18°C. After neutralisation, and a difficult separation because of the excess sodium sulphate present, the isolated disodium salt is heated with caustic soda at 350°C. This melt is then dissolved in water, neutralised, extracted using ether (a most unusual operation on the large scale) and finally distilled.

Resorcinol finds significant usage in adhesives as well as dyestuffs, with an estimated (1979) world capacity of 30 000 tonnes per annum.

Chloronitrobenzenes (VII: Figure 3.2)
Chlorobenzene is nitrated with mixed (sulphuric and nitric) acid under moderate conditions (50°C) to give a mixture of mononitro isomers, approximately 65% *para* (PCNB), 34% *ortho* (OCNB) and 1% *meta*. Separation is by a combination of fractional crystallisation and fractional distillation.

Variation of the isomer ratio is difficult to achieve and manufacturers sometimes have problems in consuming the isomers in the proportions produced. Monsanto has a patented process [10] using nitric acid in the presence of phosphoric acid and a transition metal compound as catalyst in which the *para*:*ortho* ratio is reduced from 2:1 to between 1.2:1 and 1.6:1.

Dinitrochlorobenzene (see Chapter 4) is obtained by further nitration of OCNB or PCNB, usually in an additional nitrator adjacent to the main plant to avoid extra storage and handling.

p-*Nitroaniline* (VIII: Figure 3.2)
Apart from its use as a diazo component for azo dyes, *p*-nitroaniline (PNA) is an intermediate stage in the manufacture of *p*-phenylenediamine. PNA is made by amination, i.e. heating PCNB with aqueous ammonia at 200°C under pressure, in either a batchwise or continuous operation. Conversion is high (96%) and separation easy.

p-*Phenylenediamine* (IX: Figure 3.2)
New uses for *p*-phenylenediamine (PPD) in speciality polyamide fibres such as Kevlar (DuPont) have led to re-evaluation of alternative production routes. The traditional method from PNA, employing either iron reduction or catalytic reduction, still holds sway; Figure 3.5 summarises the main production routes available. The alternatives are attractive 'paper' routes, based on cheap starting materials and having fewer stages than the established one, but they have not been scaled up to give satisfactory yield.

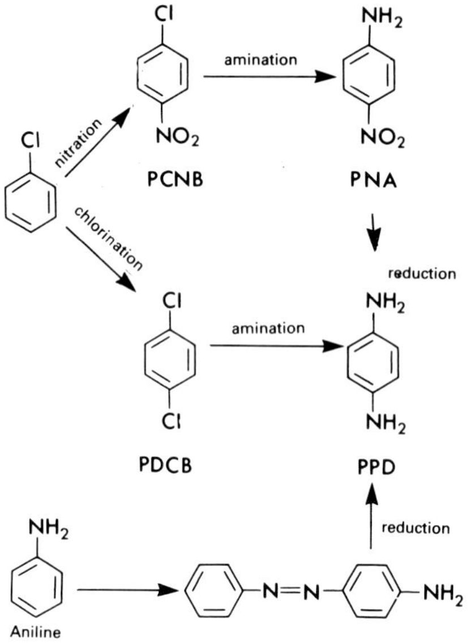

Figure 3.5 – Alternative manufacturing processes for p-*phenylenediamine*

Phenol and derivatives (X, XI, XII, XIII: Figure 3.2)
The so-called cumene process for phenol has already been referred to in Figure 3.3 and is now the dominant manufacturing method. Total world production of phenol amounts to 4 million tonnes per annum. The oxidation of cumene is complex, proceeding via the hydroperoxide and acid-catalysed cleavage. The process is economically feasible because acetone is also produced. Di-isopropylbenzene is a by-product of the cumene process,

and can be used as an alternative starting material for resorcinol and hydroquinone.

Phenol can be chlorinated using chlorine or sulphuryl chloride to form *ortho* and *para* isomers (XI: Figure 3.2), which are separated by fractional distillation. *p*-Chlorophenol is of special importance as the starting material for quinizarin (see Chapter 4).

o-Cresol (XII: Figure 3.2) is manufactured by selective methylation of phenol using methanol and a metal oxide catalyst at high temperature.

Salicylic acid (XIII: Figure 3.2) is manufactured by the Kolbe–Schmitt reaction, a specific unit process of carboxylation limited industrially to phenol and β-naphthol (cf. Chapter 4). Sodium phenate is reacted with carbon dioxide at 150°C under pressure to give a 90% conversion to salicylic acid, the major use for which is in aspirin manufacture.

p-*Aminophenol* (XIV: Figure 3.2)
Conventional processes involving iron or catalytic reduction of *p*-nitrophenol for making *p*-aminophenol are now being substituted by the potentially cheaper alternative of direct electrolytic reduction/hydroxylation of nitrobenzene. The main use of *p*-aminophenol is now in analgesic manufacture (its *N*-acetyl derivitive is paracetamol) rather than the synthesis of dyestuffs and photographic chemicals.

3.3.2 Toluene derivatives

The primaries derived from toluene are shown schematically in Figure 3.6. Names (together with initials in established usage) indicate the products that are considered in more detail.

Nitrotoluenes (XV, XVI: Figure 3.6)
Nitration of toluene is carried out under milder conditions than used for nitration of benzene to minimise side reactions. The high yield process may be batchwise or continuous and the isomer ratio is 55–60% *ortho* (ONT), 35–40% *para* (PNT) and 3–4% *meta* (cf. nitration of chlorobenzene). Separation is by a combination of fractional distillation and crystallisation.

As with chloronitrobenzene, manufacturers have developed processes that permit variation of the isomer ratio [11]. A Bofors patent [12] claims nitration in the presence of phosphoric acid raises the *para*:*ortho* ratio from 0.63:1 to 1.11:1, but commercialisation of this and many other alternatives depends on the overall economics. For example, the use of phosphoric acid would entail the scaling up of a process, calling for recovery or recycling of the acid.

2,4-Dinitrotoluene is usually manufactured by further nitration of PNT since direct nitration of toluene causes problems with dinitro isomers and other by-products [11].

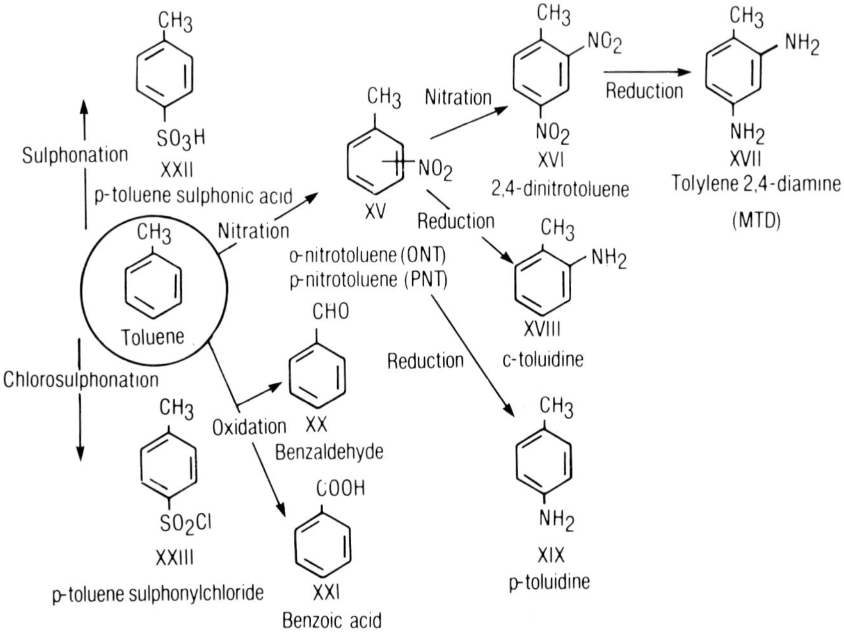

Figure 3.6 – Manufacture of primaries from toluene: industrial routes

m-*Tolylenediamine* (XVII: Figure 3.6)
Continuous catalytic hydrogenation of 2,4-dinitrotoluene is effected on the large scale using either a melt or a solvent process and, typically, a palladium/carbon catalyst. The product is purified by vacuum distillation. Most of the world production is converted, by phosgenation, to toluenediisocyanate (TDI) for polyurethane manufacture.

o-*Toluidine and* p-*toluidine* (XVIII, XIX: Figure 3.6)
Toluidines are manufactured from the corresponding nitrotoluene by reduction processes similar to those mentioned previously for aniline. Although iron reduction has been mainly superseded, catalytic reduction processes have not been developed to the same scale as those for aniline since toluidines are required in much smaller amounts.

Benzaldehyde (XX: Figure 3.6)
The classical process for the oxidation of toluene to benzaldehyde uses manganese dioxide and sulphuric acid. On the large scale it is more economic to employ catalytic oxidation. Air oxidation at 150°C with a cobalt catalyst gives a mixture with benzoic acid, whereas alternative catalysts

and conditions give a higher proportion of benzaldehyde, with benzyl alcohol as the main co-product.

The main use of benzaldehyde is as an intermediate for flavouring chemicals such as cinnamaldehyde, with lesser amounts for making triphenylmethane (TPM) dyes and certain anils.

Benzoic acid (XXI: Figure 3.6)
Air oxidation of toluene using a cobalt catalyst can give benzoic acid in high yield. The indirect route from toluene involves side-chain chlorination (vapour phase, u.v. catalysed) to benzotrichloride followed by controlled hydrolysis. This is not usually competitive with oxidation except for the direct production of benzoyl chloride (Figure 3.7).

$$PhCCl_3 + PhCOOH \xrightarrow{ZnCl_2} 2PhCOCl + HCl$$

Figure 3.7 – Benzoyl chloride manufacture from benzotrichloride

p-*Toluenesulphonic acid* (XXII: Figure 3.6)
Sulphonation of toluene with monohydrate at 100°C gives an 80% yield of the *para* isomer. The product is preferred to the simpler benzenesulphonic acid as an organic acid catalyst and in the formation of salts with coloured bases. Toluene-4-sulphonic acid is chlorinated to 2-chlorotoluene-4-sulphonic acid which, after nitration and reduction, leads to 6-chloro-3-toluidine-4-sulphonic acid.

p-*Toluenesulphonyl chloride* (XXIII: Figure 3.6)
Chlorosulphonation of toluene with excess chlorosulphonic acid at 0–20°C yields a 1:1 mixture of *ortho* and *para* isomers (lower temperatures increase the *ortho* content). These are separated by crystallisation or converted to the mixed sulphonamide and then separated. Toluene-4-sulphonyl chloride (PTSC), which may be considered as a by-product from the preparation of the *ortho* isomer required for saccharin manufacture, is used as an acylating agent (e.g. in the manufacture of esters and sulphonamides).

3.3.3 Naphthalene derivatives

The primaries derived from naphthalene are shown in Figure 3.8. Named products are those considered in more detail.

α-*Naphthylamine* (XXIV: Figure 3.8)
Nitration of naphthalene with mixed acid (HNO_3 33%, H_2SO_4 47.5%, H_2O 19.5%) can be readily controlled to give a 98% yield of α-nitronaphthalene when isolated. It is particularly important to keep the β isomer at low levels

because of the health hazards associated with the derived β-naphthylamine (manufacture of which ceased in the 1960s).

The most recent plants for the catalytic reduction of α-nitronaphthalene are specifically designed with additional safety features to minimise exposure and handling.

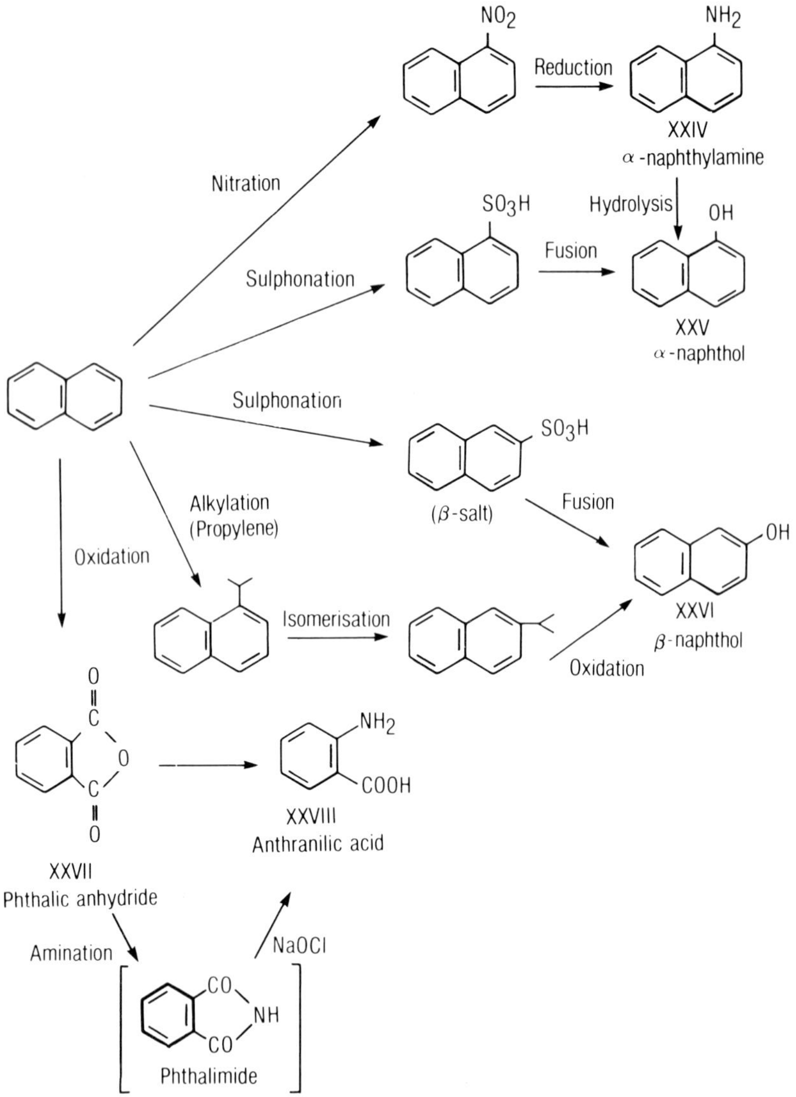

Figure 3.8 – Manufacture of primaries from naphthalene: industrial routes

α-*Naphthol* (XXV: Figure 3.8)

Sulphonation of naphthalene under mild conditions (98% H_2SO_4/40°C) forms 96% α-sulphonic acid and 4% β-sulphonic acid (β-salt). However, the α product readily isomerises to β-salt which is tedious to remove. This difficulty led to the development of the alternative route via α-naphthylamine in place of the original fusion process (the conventional terminology to describe the hydrolysis of an aromatic sulphonic acid to an aromatic hydroxy compound). The manufacturing process involves hydrolysis of α-naphthylamine with 22% sulphuric acid at 200°C in an autoclave. If high-purity product is required vacuum distillation may be necessary. A third process, giving higher-quality material directly, has been developed by Union Carbide. This involves the catalytic oxidation of tetralin to 1-tetralol/1-tetralone followed by catalytic dehydrogenation.

β-*Naphthol* (XXVI: Figure 3.8)

Sulphonation of naphthalene with 96% sulphuric acid at 165°C gives an 85% yield of the β-sulphonic acid. Separation from the minor α component is simple. The sodium salt of naphthalene-β-sulphonic acid (β-salt) is converted to β-naphthol by heating with aqueous caustic soda at 300°C for 2 h. After drowning the melt in water the β-naphthol is precipitated by adding acid and separated. The evolved sulphur dioxide is absorbed in dilute caustic soda to give sodium bisulphite solution as a by-product. The product is purified by vacuum distillation.

An alternative process, analogous to the cumene–phenol route, has been exploited by American Cyanamid [13]. Even though this alternative is essentially a three-stage process, compared with the two stages of the conventional process, the economics can be favourable provided that cheap propylene is available and the value of the by-product acetone can be realised. The key to this process is the isomerisation of the α-isopropylnaphthalene to β-isopropylnaphthalene using a solid phosphoric acid catalyst at temperatures greater than 275°C.

Phthalic anhydride (XXVII: Figure 3.8)

Most phthalic anhydride produced is used in the manufacture of plasticisers and resins. The proportion used for dyestuffs manufacture (anthraquinone, phthalocyanines and intermediates) is significant but less than 10% of a total world output of 1.5 million tonnes per annum.

Manufacture is by air oxidation of either naphthalene or *o*-xylene using vanadium-based catalysts (e.g. vanadium pentoxide on silicon with potassium sulphate as promoter). The choice of feedstock is dependent upon the relative economics of the relevant coal tar and oil-based raw materials. Process development has led to increasingly efficient and more economic processing, e.g. the utilisation of fluidised-bed in place of fixed-bed catalysis.

Anthranilic acid (XXVIII: Figure 3.8)
Anthranilic acid is derived from phthalic anhydride via phthalimide by means of the Hofmann reaction. This classical process, whereby an amide is converted into an amine with one less carbon atom by treatment with caustic soda and hypochlorite, has been developed into a continuous manufacturing process. Anthranilic acid is used in making esters and biocides, as well as in the dyestuffs industry.

3.3.4 Anthracene derivatives/anthraquinone

The major use of anthracene is the manufacture of anthraquinone, which in turn is mainly consumed in the dyestuffs industry. The importance of anthraquinone intermediates is emphasised by the fact that anthraquinone dyes, of various application classes, represent something like a quarter of the sales value of all dyestuffs produced, although a much smaller proportion in tonnage terms.

Traditional anthracene oxidation processes using oxidising agents such as dichromate or nitric acid have been supplemented by air oxidation using vanadium pentoxide catalyst. The alternative established route is based on phthalic anhydride. More recently two additional routes have been developed; all four are compared here as an example of the situation facing a manufacturer when choosing between alternative technologies for his next investment.

Figure 3.9 summarises the four alternatives. Route A tends to be operated in Europe, where purified anthracene feedstock is readily available, while Route B is more common in the USA, where anthracene is less easily obtainable but aluminium chloride is cheaper. Route B is universally operated in the manufacture of 2-methylanthraquinone and 2-chloroanthraquinone from toluene and chlorobenzene respectively. Work on the two other processes for the manufacture of anthraquinone reflects a continuing need to balance considerations of feedstock availability and total economics with process technology. Feedstock prices, the efficiency of each separate stage, processing costs and capital costs involved in building new plants are all important elements.

Route C, starting from styrene (benzene and ethylene feedstock), represents the newest technology [14]. BASF developed this process to full commercial scale and had planned to build a plant jointly with Sandoz. However, feedstock prices in the short-term have risen more rapidly than the cost of anthracene, and the route was finally abandoned because the nitration of anthraquinone from this process was shown to cause a high incidence of allergy amongst workers.

Route D has a naphthalene feedstock, and the main stage is a Diels–Alder reaction between naphthoquinone and butadiene. It has recently been selected by Bayer and Ciba–Geigy as the most appropriate technology for their new joint venture plant (Schelde Chemie), with a capacity of 12 000 tonnes per annum [15]. They claim both economic and environmental

Route A – from anthracene

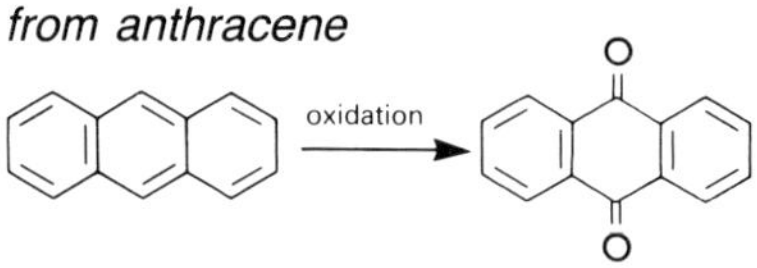

Route B – from benzene and o-xylene

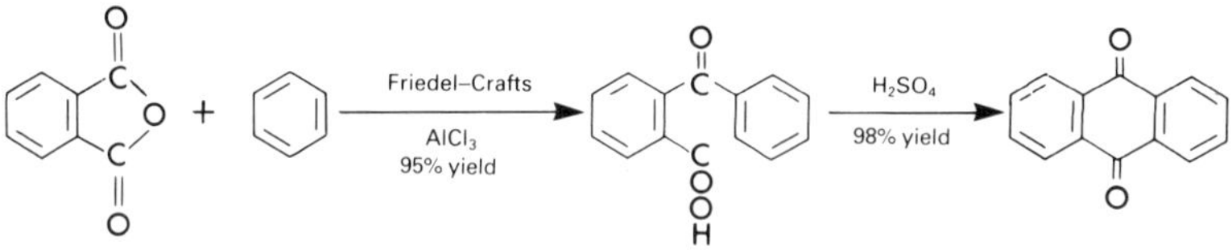

Route C – from benzene and ethylene

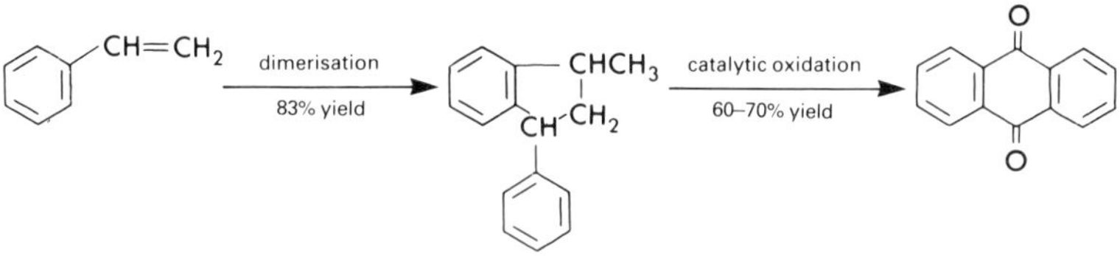

Route D – from naphthalene and butadiene

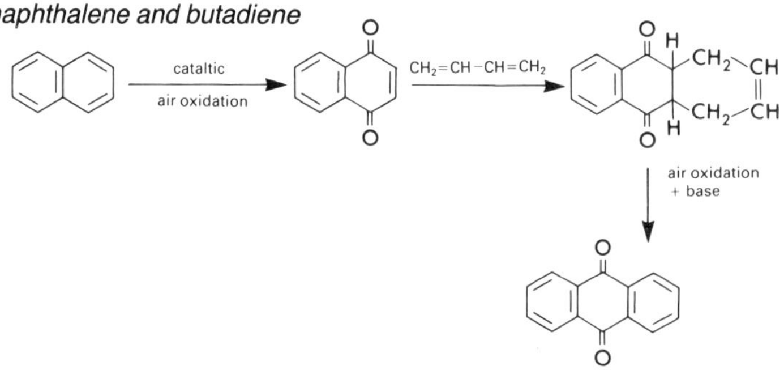

Figure 3.9 – Alternative manufacturing processes for anthraquinone

advances over other routes for their fully continuous process and must have overcome scaling up problems previously associated with this route. The economics of naphthoquinone manufacture are obviously of crucial importance and the catalytic oxidation adopted is an obvious improvement on the old IG Farben dichromate oxidation process. A major problem associated with any oxidation process is the control of the coproduced phthalic anhydride.

REFERENCES

1. Collin and Mildenberg, Chem. and Ind., (1978) 567.
2. Dierk and Stadelhofer, Chem. and Ind., (1983) 113.
3. Duncan in 'The chemical industry', Ed. Sharp and West (SCI, 1982).
4. 'Synthetic organic chemicals, US production and sales 1980' (US Government, 1981).
5. Japan Chemical Week, (Apr 1982).
6. American Cyanamid, USP 4 021 498 (1977).
7. ICI, BP 982 902/3 (1962).
8. Gans, Hydrocarbon Processing, (Nov 1976) 145.
9. Thoma (to Biller, Michaelis & Co.), USP 2 991 311 (1961).
10. Monsanto, BP 1 420 733, 1 421 107 (1973).
11. Industrial and laboratory nitrations, Ed. Albright and Hanson (ACS, 1976).
12. Bofors, BP 1 207 384 (1966).
13. American Cyanamid, USP 3 504 045/6 (1970).
14. Sherwin, Chemtech., (1974) 229.
15. Dittmar, Chem. Rundschau, (Nov 1977) 23.

CHAPTER 4

Intermediates and unit processes

4.1 DEFINITION AND SCOPE

Intermediates are defined as those precursors to colorants that are manufactured within the dyestuffs industry; they are nearly always colourless, coloured precursors being conventionally termed colour bases. As distinct from primaries they are only rarely manufactured in single-product units because of the comparatively low tonnages required. Fluorescent brightening agents (FBAs) are neither intermediates nor true colorants but, in the same way that colour bases are linked to colour manufacture, it is convenient to make any specific comments on FBA manufacture within this chapter. Of the many publications on FBAs, one of the few to deal with basic manufacturing processes is that by Zahradnik [1].

There are at least 3 000 different intermediates in current manufacture (over half that number are specifically mentioned in the *Colour Index* [2]), and in addition there is a comparatively small number of products manufactured by individual companies for their own specialities. Clearly only a selection of intermediates can be discussed here, but, since 300 of the products probably account for 90% of the quantity of intermediates used, most of the important aspects can be covered. Any meaningful quantification of world tonnage requirements of primaries and intermediates for dyes cannot readily be made.

Intermediates vary in complexity, usually related to the number of chemical and operational stages in their manufacture, and therefore cost. At 1984 price levels they may be classed as cheap (less than £1 000 per tonne, as with primaries), average (£1 000 to £3 000 per tonne) or expensive (more than £3 000 per tonne).

4.2 UNIT PROCESSES

The term unit process is used to define the chemical stage in the conversion of one chemical to another. This involves the introduction of key functional groups necessary for the conversion to the final dye (e.g. an amine group for the formation of azo dyes) or for conferring properties essential to the final product (e.g. sulphonic acid groups for water solubility). Unit processes may be categorised under at least 40 different headings [3], with 15–20 being in common use in the dyestuffs industry.

Analysis of the 30 individual primaries described in Chapter 3 (including all stages and alternative processes) shows the occurrence of ten different unit processes. These are tabulated (Table 4.1) to compare the pattern of usage and give some impression of the relative importance. Table 4.1 additionally lists other unit processes for further discussion. The various groups required as substituents in the final product (e.g. nitro, hydroxyl or sulphonic acid) are seen to be predominantly introduced at the intermediate stage, rather than when reacting a coloured product. Exceptions to this would be with certain anthraquinone products and the halogenation of copper phthalocyanine.

TABLE 4.1

Unit processes in dyestuff manufacture

Process	*Primaries (No. of occurrences within 30 identified product manufactures)*	*Intermediates (common usage)*	*Colorants (common usage)*
Nitration	6	√	
Reduction	8	√	
Sulphonation	4	√ (incl. chloro sulphonation)	
Oxidation	5	√	
Fusion/hydroxylation	3	√	
Amination	3	√ (incl. Bucherer reaction)	
Alkylation	2	√	
Halogenation	2	√	√
Hydrolysis	2	√	
Condensation	1	√	√
Alkoxylation		√	
Esterification		√	
Carboxylation		√	
Acylation		√	√
Phosgenation		√	√
Diazotisation		√	√
Coupling (azo)			√

Basic routes for the production of intermediates have remained essentially the same over the last 50 years. In this area the authoritative reviews of Fierz-David [4] and Venkataraman [5] remain relevant. More efficient processes have since been developed, but most of the change has been brought about by major improvements in the field of unit operations (Chapter 6). The classic source book on unit processes is by Groggins [6], and it is interesting to use this to review progress over the last 25 years.

Figures 4.1 to 4.7 summarise some of the major chemical routes, accentuating the unit processes involved, which derive from the key primaries

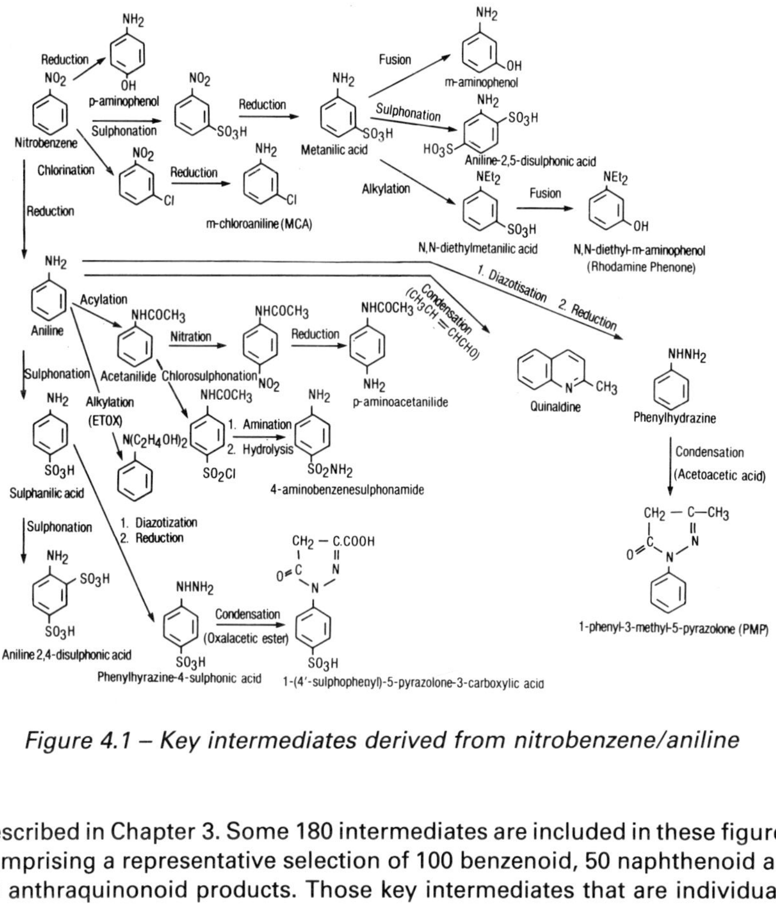

Figure 4.1 – Key intermediates derived from nitrobenzene/aniline

described in Chapter 3. Some 180 intermediates are included in these figures, comprising a representative selection of 100 benzenoid, 50 naphthenoid and 30 anthraquinonoid products. Those key intermediates that are individually named (often with the trivial name in common usage rather than the formal name evident from the formula) are of sufficient importance to be available for purchase throughout the world. Dyestuff manufacturers have the choice of purchasing intermediates or manufacturing them, and very few companies are totally self-sufficient in intermediates. Many of the examples from these seven figures will be used later in separate discussion of the key unit processes as listed in Table 4.1

Figure 4.1 demonstrates how two primaries, when subjected to a variety of unit processes (in addition to the formation of derived primaries as shown in Figure 3.2) give rise to ten important substituted anilines and several *N*-heterocyclic derivatives.

Figures 4.2 and 4.3 show how families of products arise from OCNB and PCNB, the most versatile of all primaries. By employing combinations of all the important unit processes essential groups of diazo components such as nitroanilines, anisidines and aminophenols are manufactured. The longest

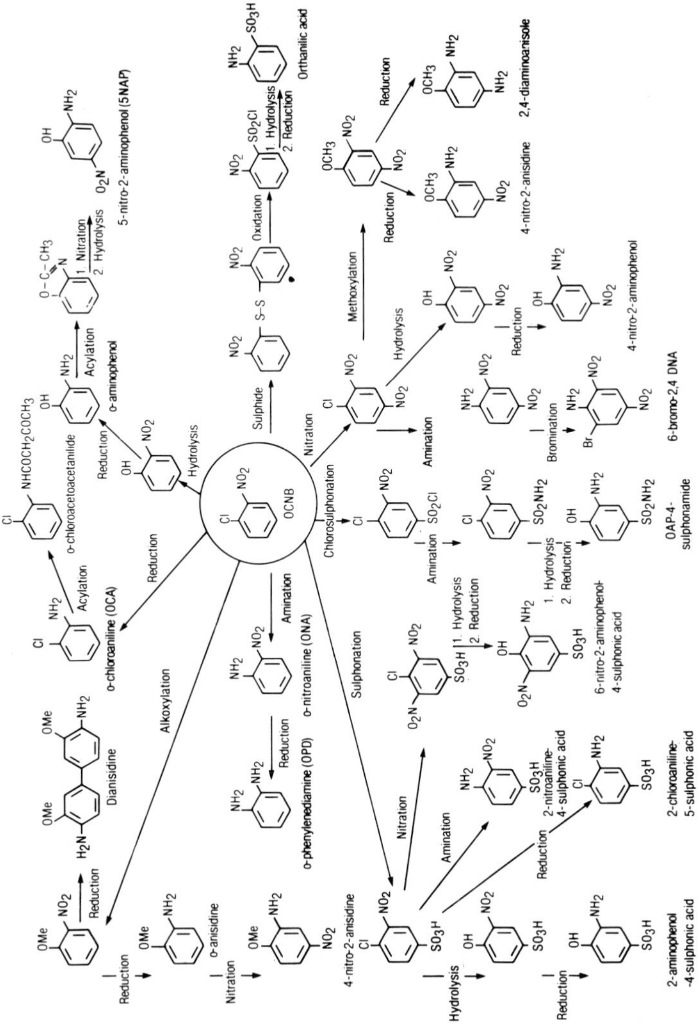

Figure 4.2 – Key intermediates derived from o-chloronitrobenzene

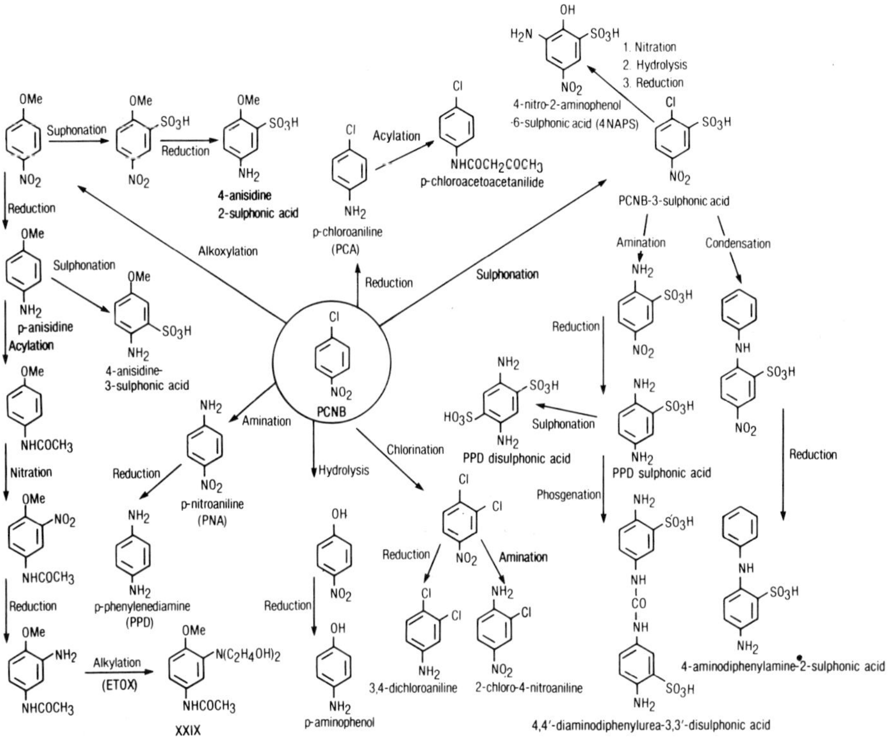

Figure 4.3 – Key intermediates derived from p-*chloronitrobenzene*

sequence of six stages is highlighted in Figure 4.3 as a typical process to form a disperse dye coupling component (XXIX). The complexity of this six-stage intermediate combined with a three-stage diazo component such as 6-bromo-2,4-dinitroaniline (Figure 4.2) provides an example of a typical navy azo disperse dye requiring ten process stages.

The three nitrotoluene isomers and the derived toluidines are further reacted to give the wide range of intermediate types shown in Figure 4.4. They include diazo and coupling components together with the very important 4,4′-diaminostilbene-2,2′-disulphonic acid intermediate for water-soluble FBAs.

Figures 4.5 and 4.6 show that naphthalene intermediates are predominantly sulphonated products. Most of them are naphtholsulphonic acids, naphthylaminesulphonic acids or aminonaphtholsulphonic acids that are important coupling components for direct and acid dyes. The majority of the products are formed by a combination of the following unit processes: sulphonation, nitration, reduction, amination and fusion, in various sequences.

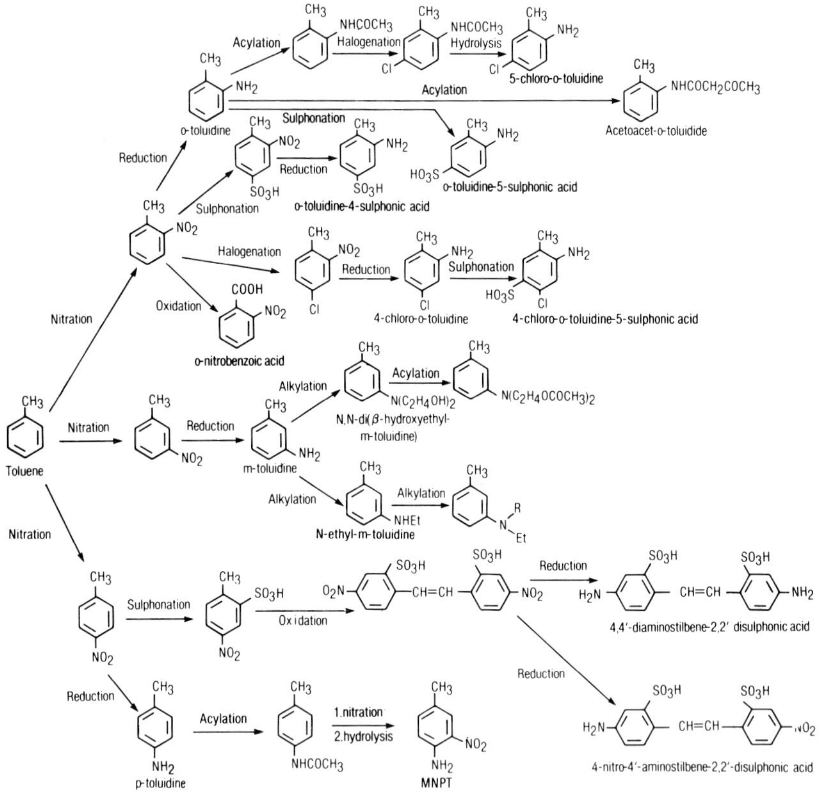

Figure 4.4 – Key intermediates derived from toluene primaries

In Figure 4.6 are the two isomeric monosulphonic acids, two of the ten possible isomers of naphthalenedisulphonic acids and two of the 14 possible trisulphonic acids. Much more information on isomer formation is available in the specialist work by Donaldson [7], but the importance of precise reaction conditions and isomer separation is obvious and will be discussed further (Section 4.5).

A high proportion of anthraquinone intermediates are formed by sulphonation, nitration, fusion and amination, together with halogenation, as shown in Figure 4.7. But, as will be discussed later, the conditions employed are significantly different from the naphthalene series and reactions often require catalysing. An important initial comparison brought out in Figure 4.7 is with alternative routes to α-substituted and 1,5- and 1,8-disubstituted products, based on either α-sulphonation or α-nitration.

The importance of route selection for primaries has already been emphasised in Chapter 3. The situation with intermediate manufacture is frequently

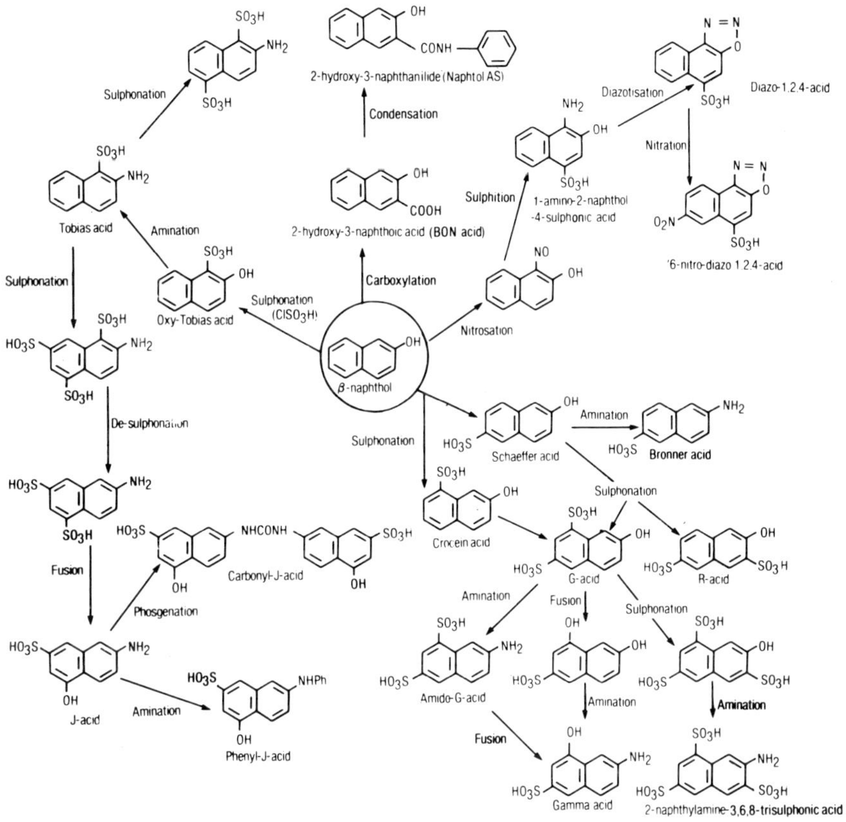

Figure 4.5 – Key intermediates derived from β-naphthol

similar, with several factors governing the choice of route, and these are not wholly directed at the cost of the final product. Examples of this will be brought out later by comparisons of unit process options.

With such a wide range of processing facilities available, manufacture of new intermediates can usually be sited in existing plants. Novel aromatic heterocyclic derivatives, such as substituted pyridones and thiophenes, are in this category and represent a trend towards the use of more sophisticated intermediates to obtain special effects.

4.3 NITRATION

The unit process of nitration is concerned with the introduction of one or more nitro (NO_2) groups into an aromatic nucleus by the replacement of a hydrogen atom. For all the more important intermediates to be considered

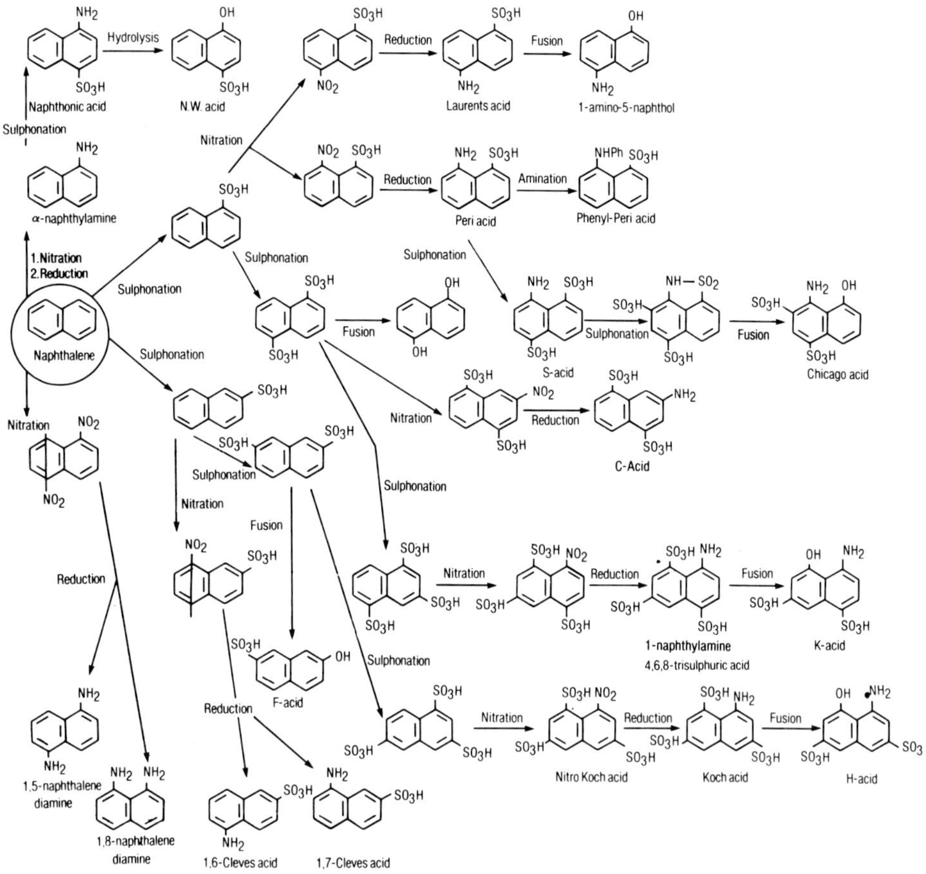

Figure 4.6 – Key intermediates derived from naphthalene

the reaction is achieved by the addition of nitric acid, usually in combination with sulphuric acid (mixed acid), to a sulphuric acid solution of the reactant. The nitronium ion (NO_2^+) is the active nitrating species and the key process variables that affect yield and quality of product are concentration, temperature and mixed acid composition.

The nitro product is obtained by drowning out (i.e. adding to excess water with internal or external cooling as required) followed by separation or extraction.

The largest scale nitrations, already described in Chapter 3, are those where the nitro compounds are typically precursors of amine primaries. When nitrating benzene, toluene, chlorobenzene and naphthalene continuous operation in custom-built plants can normally be justified. These primary

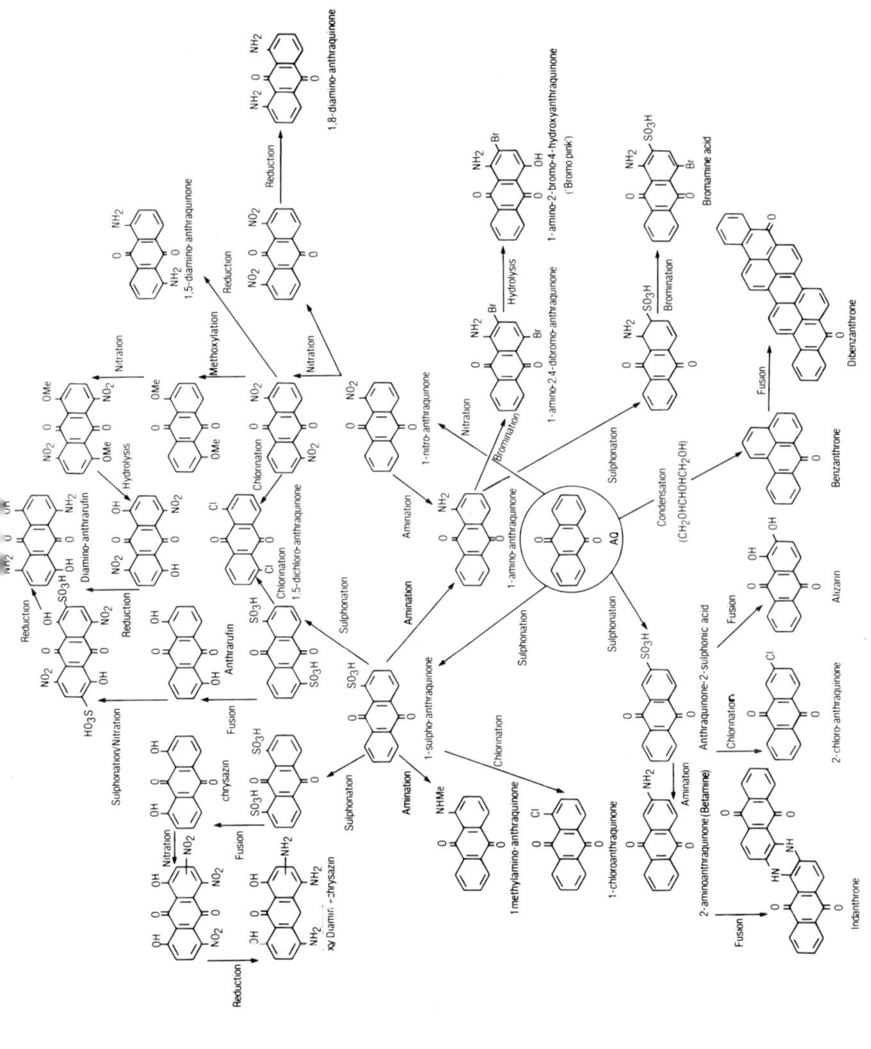

Figure 4.7 – Key intermediates derived from anthraquinone

nitro products are also typical of the isomer patterns and separation processes that are paralleled in more highly substituted molecules. Apart from the normal economic advantages of continuous operation (i.e. lower operating expenses, including energy, and capital investment per unit of production), continuous nitration offers better controlled conditions, which together with a lower in-process inventory lead to safer operation.

The remaining smaller-scale nitrations are typically carried out in multi-product nitration units. Occasionally nitration is combined with sulphonation to give a telescoped two-stage reaction in a single unit.

A typical batch nitration plant arrangement is shown in Figure 4.8 (cf. Figure 3.4 for continuous nitration plant). The reaction vessel (nitrator) is constructed of cast iron, mild carbon steel or stainless steel depending upon the reaction environment. It is designed to maintain the required operating temperature with heat-removal capability to cope with this strongly exothermic and potentially hazardous (see Section 7.3) reaction.

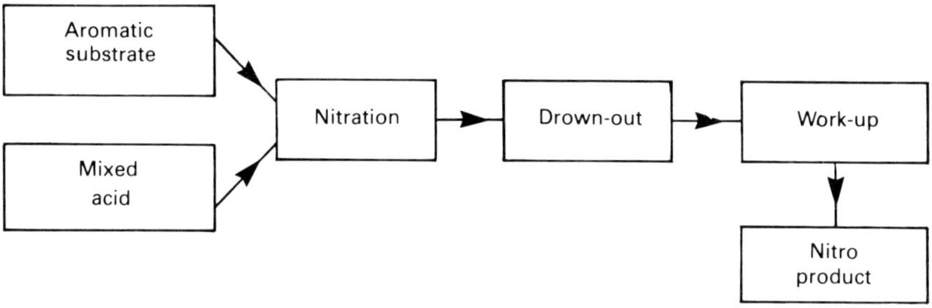

Figure 4.8 – Arrangement of nitration plant

Secondary problems are the containment of nitric oxide fumes and disposal or reuse of the dilute 'spent' acid. Examples of important intermediates resulting from nitration are summarised in Table 4.2. The relevance of most of these intermediates may be derived from Figures 4.1 to 4.7.

TABLE 4.2

Intermediates obtained by nitration

Product	*Starting material*	*Figure*	*Nitration process*
1,2-Dichloro-4-nitrobenzene	*o*-Dichlorobenzene		Mixed acid/30°C and purification from isomers
1,4-Dichloro-2-nitrobenzene	*p*-Dichlorobenzene		Mixed acid/30°C
4-Chloro-2-nitrotoluene	*p*-Chlorotoluene		Mixed acid/20°C and fractionated from 35% 4-chloro-3-nitrotoluene

Table 4.2

Product	*Starting material*	*Figure*	*Nitration process*
2-Nitro-1,4-xylene	*p*-Xylene		Mixed acid/20°C
4-Nitro-1,3-xylene	*m*-Xylene		Mixed acid/25°C 2-nitro (13% yield) distilled off from 4-nitro (72% yield)
Nitromesitylene	Mesitylene		Mixed acid/0°C
1,5/1,8-Dinitronaphthalene	Naphthalene	4.6	Mixed acid/80°C (50% yield 1,8 + 25% 1,5 after separation)
2,6-Dinitrochlorobenzene-4-sulphonic acid	OCNB sulphonic acid	4.2	Mixed acid/100°C
2,4-Dinitrochlorobenzene-6-sulphonic acid	PCNB sulphonic acid	4.3	Mixed acid/100°C
5/8-Nitronaphthalene-1-sulphonic acid	Naphthalene-α-sulphonic acid	4.6	HNO_3/30°C 40:60 mix 1,5 to 1,8
5/8-Nitronaphthalene 2-sulphonic acid	Naphthalene-β-sulphonic acid	4.6	HNO_3 (50:50 mix)
3-Nitronaphthalene-1,5-disulphonic acid	Naphthalene-1,5-disulphonic acid	4.6	HNO_3/oleum to minimise 4-nitro
1-Nitronaphthalene-4,6,8-trisulphonic acid	Naphthalene-1,3,5-trisulphonic acid	4.6	Mixed acid/30°C through process from naphthalene
1-Nitronaphthalene-3,6,8-trisulphonic acid (nitro-Koch)	Naphthalene-1,3,6-trisulphonic acid	4.6	Mixed acid/40°C through process
4-Nitroacetanilide	Acetanilide	4.1	Mixed acid/0–5°C (90% yield)
3-Nitro-*p*-toluidine	*N*-Acetyl-*p*-toluidine	4.4	Mixed acid/25°C and hydrolysis
4-Nitro-2-anisidine	(a) *o*-Anisidine	4.2	
	(b) *o*-Anidisine nitrate		Low temp. rearrangement
2-Nitro-*N*-acetyl-*p*-anisidine	*N*-Acetyl-*p*-anisidine	4.3	Mixed acid/25°C
6-Nitrodiazo-1,2,4-acid	Diazo-1,2,4-acid	4.5	Mixed acid/5°C (95% yield)
1-Nitroanthraquinone	Anthraquinone	4.7	HNO_3/85% sulphuric acid
1,5/1,8-Dinitroanthraquinone	1-Nitroanthraquinone	4.7	HNO_3/monohydrate/100°C separation by fractional crystallisation
4,8-Dinitroanthrarufin-2,6-disulphonic acid	Anthrarufin	4.7	Oleum sulphonation followed by mixed acid/30°C
4,5-Dinitrochrysazin (+ isomers)	Chrysazin	4.7	HNO_3/oleum/0°C
5-Nitro-2-aminophenol	2-Methylbenzoxazole	4.2	Mixed acid/0–10°C hydrolysis and separation from 4-nitro
5-Nitro-2-aminothiazole	2-Aminothiazole nitrate		Low temp. rearrangement

Very few products have alternative nitration processes that are economically viable. 4-Nitro-2-anisidine is an unusual example in being available either by carefully controlled nitration of 2-anisidine (note that most arylamines require protection, e.g. by *N*-acetylation, prior to nitration) or by adding pre-formed 2-anisidine nitrate to sulphuric acid [8]. Manufacture of 5-nitro-2-aminothiazole is the only other example of this nitrate rearrangement process starting from 2-aminothiazole nitrate.

A number of products can be made by alternative production routes, for either the nitro compound or the derived amine:

(a) 4-Nitro-2-anisidine can be manufactured by partial reduction of 2,4-dinitroanisole (Figure 4.2)
(b) 4-Chloro-2-nitrotoluene can also be obtained by chlorination of 2-nitrotoluene (Figure 4.4), although there is still an isomer separation problem
(c) *m*-Chloronitrobenzene is similarly obtained by chlorination of nitrobenzene to supplement material produced as the minor (1%) isomer from the nitration of chlorobenzene (see Section 3.3.1)
(d) α-Nitronaphthalene competes with naphthalene-α-sulphonic acid as a route to α-naphthol (see Section 3.3.3)
(e) 1-Nitroanthraquinone and anthraquinone-1-sulphonic acid are both used as alternative intermediates for producing 1-aminoanthraquinone (Figure 4.7)
(f) 1,5- and 1,8-Dinitroanthraquinones can similarly be compared as alternative intermediates to 1,5- and 1,8-anthraquinonesulphonic acids in the manufacture of the important diaminoanthraquinone derivatives (Figure 4.7).

The importance of the difficult α-nitration of anthraquinone can be assessed by the development work taking place across the world reflected by well over 100 process patents published during the last 15 years. Of major importance is the adoption of α-nitration in place of α-sulphonation by Bayer in its new Schelde-Chemie facility, which has a quoted capacity of 10 000 tonnes per year of 1-nitroanthraquinone and 8 000 tonnes per year of disubstituted derivatives [9]. Detailed nitration kinetics and isomer variation has also been reported by Sandoz [10], leading to design of a pilot plant consisting of a stirred tank overflowing through a tubular reactor at 30°C.

4.4 REDUCTION

4.4.1 Arylamines

The most important reduction process is the conversion of an aromatic nitro or dinitro compound into an arylamine or arylene diamine. This unit process may be alternatively called amination by reduction [6]. The six hydrogen atoms per nitro group that are required ($RNO_2 + 6H \rightarrow RNH_2 + 2H_2O$) are provided either directly, by catalytic hydrogenation, or indirectly using a wide range of reagents and operating conditions.

The following types of reduction are all exemplified in Table 4.3, which lists the industrial processes for intermediates mentioned earlier in this chapter together with primaries from Chapter 3:

(a) Hydrogen/catalyst/liquid feed and product
(b) Hydrogen/catalyst/solvent
(c) Hydrogen/catalyst/aqueous
(d) Neutral iron (aqueous)
(e) Acidic iron (aqueous)
(f) Iron/solvent
(g) Sulphide used for (i) selective reduction of dinitro compounds to nitroarylamines or (ii) metal-sensitive systems such as certain substituted 2-aminophenols.

TABLE 4.3

Arylamines obtained by reduction

Product	*Starting material*	*Figure*	*Reduction process*
Aniline	Nitrobenzene	3.3	Catalytic hydrogenation
p-Aminophenol	Nitrobenzene	3.2	Electrolytic
m-Phenylenediamine	*m*-Dinitrobenzene	3.2	Catalytic hydrogenation
m-Nitroaniline	*m*-Dinitrobenzene		Sulphide
o-Chloroaniline	*o*-Chloronitro-benzene	4.2	Iron
m-Chloroaniline	*m*-Chloronitro-benzene	4.1	Iron
p-Chloroaniline	*p*-Chloronitro-benzene	4.3	Iron
3,4-Dichloroaniline	3,4-Dichloronitro-benzene	4.3	Iron
2,5-Dichloroaniline	2-Nitro-1,4-dichlorobenzene		Iron
o-Toluidine	*o*-Nitrotoluene	3.6	Catalytic hydrogenation
m-Toluidine	*m*-Nitrotoluene	4.4	Iron or catalytic hydrogenation
p-Toluidine	*p*-Nitrotoluene	3.6	Catalytic hydrogenation
4-Chloro-*o*-toluidine	4-Chloro-2-nitro-toluene	4.4	Iron
m-4-Xylidine	4-Nitro-1,3-xylene		Iron
Mesidine	Nitromesitylene		Iron
α-Naphthylamine	α-Nitronaphthalene	3.7	Catalytic hydrogenation
1,5-Naphthylene diamine	1,5-Dinitro-naphthalene	4.6	Iron/acetic acid or catalytic hydrogenation
1,8-Naphthylene diamine	1,8-Dinitro-naphthalene	4.6	Catalytic hydrogenation
o-Aminophenol	*o*-Nitrophenol	4.2	Sulphide
4-Nitro-2-aminophenol	2,4-Dinitrophenol	4.2	Sulphide/ammonia
4-Chloro-2-aminophenol	4-Chloro-2-nitrophenol		Iron/H_2SO_4
6-Chloro-4-nitro-2-aminophenol	6-Chloro-2,4-dinitrophenol		Sulphide
4-Chloro-6-nitro-2-aminophenol	4-Chloro-2,6-dinitrophenol		Sulphide

Table 4.3

Product	*Starting material*	*Figure*	*Reduction process*
o-Anisidine	*o*-Nitroanisole	4.2	Iron/formic acid
p-Anisidine	*p*-Nitroanisole	4.3	Sulphide/115°C
4-Nitro-2-anisidine	2,4-Dinitroanisole	4.2	Sulphide
2,4-Diaminoanisole	2,4-Dinitroanisole	4.2	Catalytic hydrogenation
Orthanilic acid	*o*-Nitrobenzene-sulphonic acid	4.2	Iron
Metanilic acid	*m*-Nitrobenzene-sulphonic acid	4.1	Iron or catalytic hydrogenation
2-Chloroaniline-5-sulphonic acid	2-Chloronitrobenzene-5-sulphonic acid	4.2	Iron/HCl
o-Toluidine-4-sulphonic acid	*o*-Nitrotoluene-4-sulphonic acid	4.4	Iron
3,5-Diaminomesitylene sulphonic acid	3,5-Dinitro-mesitylene sulphonic acid		Iron
Laurent's acid	1-Nitronaphthalene-5-sulphonic acid	4.6	Iron } may be separated after reduction of mixed nitro isomers
Peri acid	1-Nitronaphthalene-8-sulphonic acid	4.6	Iron
2-Naphthylamine-4,8-disulphonic acid	3-Nitronaphthalene-1,5-disulphonic acid	4.6	Iron
1-Naphthylamine-4,6,8-trisulphonic acid	8-Nitronaphthalene-1,3,5-trisulphonic acid	4.6	Iron
Koch acid	8-Nitronaphthalene-1,3,6-trisulphonic acid	4.6	Iron
1,6- and 1,7-Cleves acid	5/8-Nitronaphthalene-2-sulphonic acid	4.6	Iron
2-Aminophenol-4-sulphonic acid	2-Nitrophenol-4-sulphonic acid	4.2	Sulphide
2-Aminophenol-4-sulphonamide	2-Nitrophenol-4-sulphonamide	4.2	Iron/acid
6-Nitro-2-aminophenol-4-sulphonic acid	2,6-Dinitrophenol-4-sulphonic acid	4.2	Sulphide
4-Nitro-2-aminophenol-6-sulphonic acid	2,4-Dinitrophenol-6-sulphonic acid	4.3	Sulphide
4-Anisidine-2-sulphonic acid	2-Methoxy-5-nitro-benzenesulphonic acid	4.3	Iron
p-Phenylenediamine	*p*-Nitroaniline	3.5	Iron (90% yield) or catalytic hydrogenation
p-Aminoacetanilide	*p*-Nitroacetanilide	4.1	Iron
3-Amino-4-methoxy-acetanilide	3-Nitro-4-methoxy-acetanilide	4.3	Iron
p-Phenylenediamine-4-sulphonic acid	4-Nitroanaline-2-sulphonic acid	4.3	Iron
4-Aminodiphenylamine-2-sulphonic acid	4-Nitrodiphenylamine-2-sulphonic acid	4.3	Iron
4-Nitro-4′-amino-stilbene-2,2′-disulphonic acid	4,4′-Dinitrostilbene-2,2′-disulphonic acid	4.4	Sulphide

Table 4.3

Product	*Starting material*	*Figure*	*Reduction process*
4,4′-Diaminostilbene-2,2′-disulphonic acid	4,4′-Dinitrostilbene-2,2′-disulphonic acid	4.4	Iron
1-Aminoanthraquinone	1-Nitroanthraquinone	4.7	Sulphide
1,5-Diamino-anthraquinone	1,5-Dinitro-anthraquinone	4.7	Sulphide (or NH_3 amination)
1,8-Diamino-anthraquinone	1,8-Dinitro-anthraquinone	4.7	
4,5-Diamino-chrysazin (crude)	*x,y*-Dinitro-chrysazin	4.7	Sulphide
4,8-Diamino anthrarufin-2,6-disulphonic acid	4,8-Dinitroanthrarufin-2,6-disulphonic acid	4.7	Sulphide

This comparison of intermediates with primaries is especially important since it should be clear that virtually all arylamines now manufactured by iron reduction are capable of being manufactured by catalytic hydrogenation when the transition is justified by demand. Where both iron reduction and hydrogenation are ascribed processes both are known to have been used.

Catalytic hydrogenation can be represented by the plant arrangement outlined in Figure 4.9. The reducer itself is usually a stirred pressure vessel with a configuration to suit the need of the particular type of process employed. An important variant is the loop reactor, which eliminates agitator sealing problems and improves heat transfer. Faster reaction times and lower catalyst consumption are claimed (Buss AG, Basel) for certain arylamines.

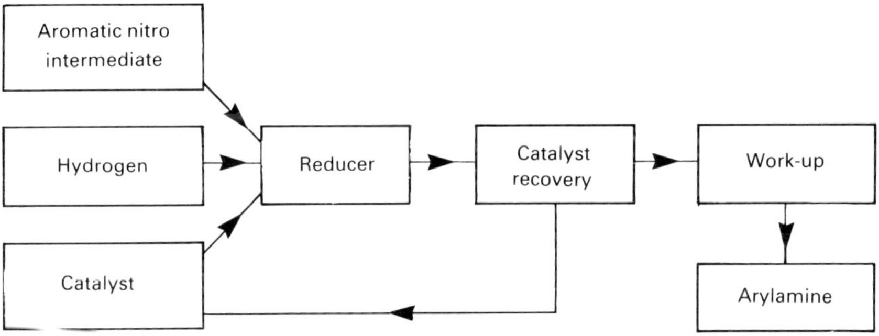

Figure 4.9 – Arrangement of plant for manufacture of arylamines by catalytic hydrogenation

The first considerations are the physical form and solubility in water or in solvent of the starting material and product. Since the catalyst has to be separated, and ideally recycled, before final work-up towards a product of required purity this normally implies a screening (filtration) stage from a fluid solution. Aqueous hydrogenations are normally limited to sulphonated species, but even here arylamine sulphonic acids often have limited solubility. It is sometimes feasible to have a non-solvent process when the product is an oil amenable to filtration. The most universally applicable process is to start with the nitro compound in a suitable solvent then separate the catalyst and work-up the arylamine by removal of solvent, possibly followed by distillation. A wide variety of metal and supported metal catalysts are used, with nickel and palladium predominating.

Operating costs for hydrogenation are cheaper than those for iron or sulphide reduction, especially when the separation and disposal of the iron oxide sludge or the sulphide liquors, are taken into account. The catalyst cost makes only a small contribution, and with built-in recovery it is insignificant. The principal reason for lack of universal application is the cost of hydrogen if it has to be transported and used from cylinders. The long-term solutions to this problem are either to move the hydrogenation to the source of hydrogen or the development of a cheaper source of local hydrogen, e.g. an electrolytic cell.

Although it is difficult to generalise about iron reduction processes, they are all carried out in cast iron vessels (2 000 to 12 000 l) with especially robust and powerful agitators to cope with the dense charge of iron borings. Aqueous acid predominates as the reaction medium, with sulphuric or hydrochloric acid or the cheaper 'pickle liquor' normally being used. In specific instances the soluble iron content is enhanced by the addition of a ferrous salt, and occasionally better results accrue from the use of formic or acetic acids in place of an inorganic acid.

Sometimes better results are obtained when the acid is omitted, and in such cases the reactions are referred to as neutral iron reductions. In all cases the quality and size (varying from fine powder to 60 mesh borings) of the iron is important. Where the nitro compound or the amine product has insufficient water solubility the process is modified by the introduction of a solvent (e.g. ethanol), but with additional cost and hazard penalties.

Whatever the process variant, the nitro compound is added slowly to the iron suspension at 80–100°C. Reaction is rapid and the nitro compound is reduced quickly and exothermically. Build-up of unreduced excess nitro compound must be avoided and the final mixture tested for its total absence. A typical work-up involves basification with soda ash (anhydrous sodium carbonate) to remove soluble iron, and filtration from the co-produced iron oxides. Liquid products are separated or solid products isolated by filtration after acidification and/or salting of the aqueous filtrate. With solvent reductions isolation is preceded by a solvent evaporation stage. If the product is a steam-volatile oil then a convenient alternative separation is to steam distil from the iron reduction mixture following basification. After optimising the many process variables, isolated yields are high at around 90%.

For alkali sulphide reductions there is a choice between the reagents sodium hydrosulphide and sodium sulphide. Sulphide reductions are advantageous and are therefore generally applied to (a) the manufacture of nitroamines from dinitro compounds, (b) the reduction of nitrophenols, (c) the reduction of nitroanthraquinones and (d) the manufacture of aminoazo compounds from the corresponding nitroazo derivative. The reactions are slow and usually achieved by controlled addition of the nitro compound to the dilute aqueous sulphide at 80–90°C. For mononitro compounds excess sulphide may be employed, but for selective reduction of a dinitro or nitroazo compound then the molecular proportion of sulphide is critical. Yields are between 70 and 90% and product isolation is usually achieved by cooling and salting out.

A number of intermediates listed in Table 4.3 have alternative manufacturing routes:

(a) 3,4-Dichloroaniline by selective amination of 1,2,4-trichlorobenzene
(b) *m*-Toluidine by high temperature amination of *m*-cresol
(c) 4-Nitro-2-anisidine by nitration of *o*-anisidine (Table 4.2)
(d) 1,5-Diaminoanthraquinone by amination of anthraquinone-1,5-disulphonic acid
(e) 4,8-Diaminoanthrarufin by desulphonation of the 2,6-disulphonic acid using hydros, but for the pure product a more specific process is employed: 1,5-diphenoxyanthraquinone (Section 4.13) is nitrated to give a hexanitro derivative, which on hydrolysis gives 4,8-dinitroanthrarufin for subsequent sulphide reduction after separation of the by-product, 2,4-dinitrophenol.

4.4.2 Benzidine and similar rearrangements

Although manufacture of benzidine itself has virtually ceased in Europe and the USA (Section 7.4), parallel rearrangement processes are operated for 3,3′-dichlorobenzidine, *o*-dianisidine (3,3′-dimethoxybenzidine (Figure 4.2)) and benzidine-2,2′-disulphonic acid (from nitrobenzene-3-sulphonic acid). The nitro compound (exemplified for OCNB in Figure 4.10) is reduced either by zinc and alkali or electrolytically to the hydrazo intermediate, which is then transformed to the benzidine form by treatment with acid.

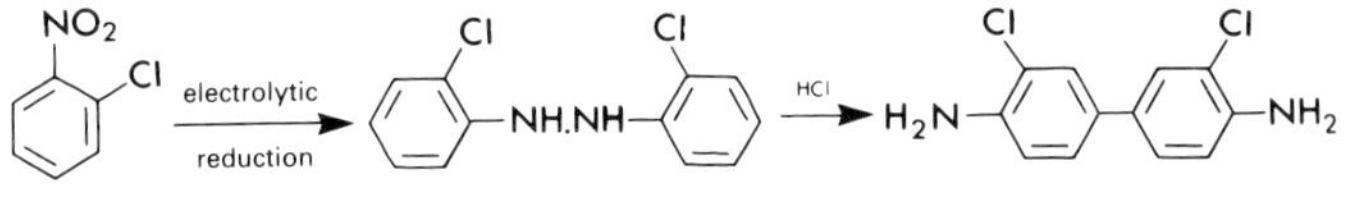

Figure 4.10 – Manufacture of 3,3′-dichlorobenzidine from o*-chloronitrobenzene*

4.4.3 Arylhydrazines

Another important reduction process is the reaction of aryldiazonium salts with sulphite/bisulphite at controlled pH to produce arylhydrazines by a four-stage process, as typified for phenylhydrazine in Figure 4.11. Arylhydrazines, as a class, are important intermediates for the preparation of pyrazolones

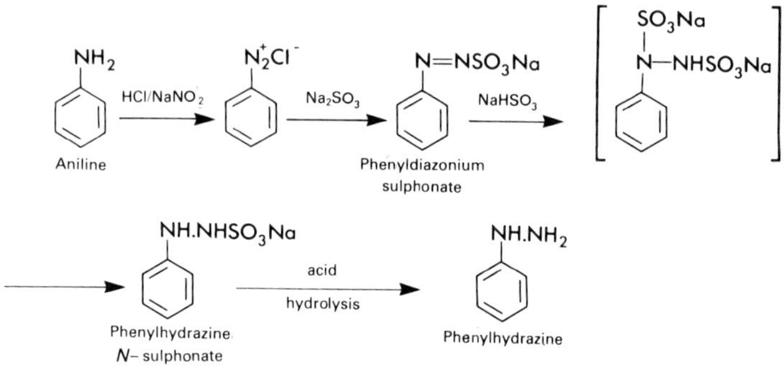

Figure 4.11 – Manufacture of phenylhydrazine from aniline

and indoles (Section 4.12). Similar aqueous processes are operated starting from a wide range of arylamines, including toluidines, nitroanilines, chloroanilines, metanilic acid, sulphanilic acid (Figure 4.1) and 2,5-dichloroaniline-4-sulphonic acid.

4.4.4 Quinone reductions

In the benzene and naphthalene series there are few examples of quinone reductions other than of hydroquinone itself (Section 3.3.1). There are, however, many intermediate chains in the anthraquinone series that depend on the generation, usually by employing aqueous hydros, of the so-called leuco compound. The reaction with leuco quinizarin is shown in Figure 4.12 since this provides the key route to the important 1,4-diaminoanthraquinones [11].

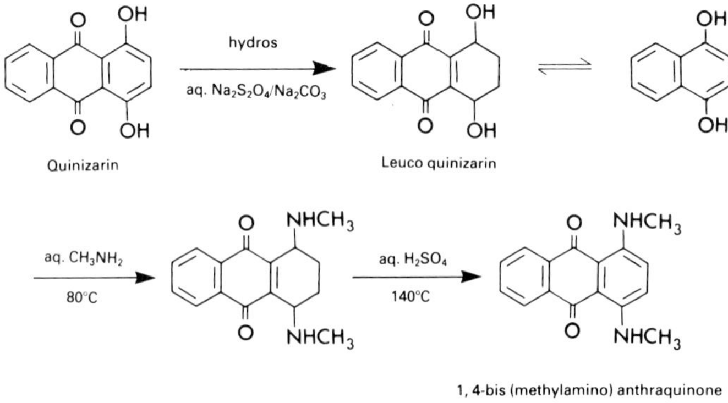

Figure 4.12 – Typical reaction with leuco quinizarin

4.4.5 Double bond reductions

Aliphatic carbon double bond reductions are rare in the manufacture of intermediates, and indeed aromatic double bond reductions too are not common. Cyclohexylamine (which should really be considered as a primary) and substituted cyclohexylamines are obtained by hydrogenation (cobalt-based catalyst with alkaline promoter) of the corresponding aniline.

Occasionally azo double bond reductions are carried out as a means of specific introduction of an amino group into the normal coupling position of a typical coupling component. Apart from the example of *p*-phenylenediamine from 4-aminoazobenzene (Section 3.3.1), there is 5-aminosalicylic acid obtained from the sulphide reduction of [aniline → salicylic acid] and 4-amino-1-naphthol obtained by hydros reduction of [sulphanilic acid → α-naphthol].

4.4.6 Sulphinic acids and mercaptans

Arylalkylsulphones are important intermediates obtained by alkylation (Section 4.9) of arylsulphinic acids. The latter are obtained by reduction of the corresponding sulphonyl chloride (Section 4.5). This reduction process is simple and of general application involving the addition of the isolated sulphonyl chloride paste to excess aqueous sodium sulphite followed by salting-out the product and isolation.

With more rigorous reduction conditions, such as zinc/acid, sulphonyl chlorides are reduced through to arylmercaptans. Examples such as 2-chloro-5-mercapto-*p*-xylene and thio-β-naphthol are manufactured from the corresponding sulphonyl chlorides as thioindigoid precursors.

4.5 SULPHONATION

The introduction of the sulphonic acid ($-SO_3H$) group into an aromatic nucleus is achieved by a variety of sulphonating agents and a wide range of processes. The large number of sulpho-containing intermediates is due to the need for ranges of water-soluble dyes (e.g. direct and acids), with the sulphonic acid being predominant as the solubilising group.

The reaction, like many other electrophilic substitutions, is reversible and the group being introduced has therefore the potential to migrate around the aromatic nucleus. This reversibility and migration is most frequently found with naphthalene substitution, and the desulphonation of naphthalene-α-sulphonic acid derivatives is frequently exploited to obtain the specific isomer required, e.g. 2-naphthylamine-5,7-disulphonic acid (Figure 4.5).

A typical batch sulphonation plant arrangement is shown in Figure 4.13. The reaction vessel (sulphonator) is constructed of cast iron with a capacity in the range 1 000 to 10 000 l. The drown-out vessel requires a lining (tile, lead or rubber) to cope with the wide range of acid concentrations and temperatures encountered. The acid filtrate after isolation is generally too

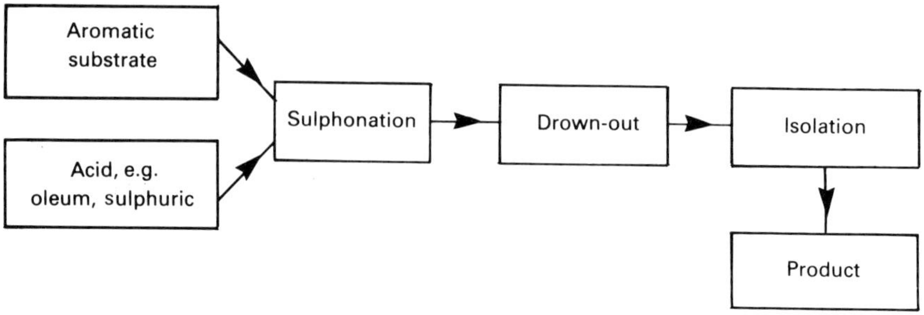

Figure 4.13 – Arrangement of sulphonation plant

dilute for economical recovery, so it requires neutralisation with lime or waste alkali before disposal.

All the following sulphonation processes are used to manufacture sulphonic acids and the examples specified in Table 4.4 will serve to demonstrate comparisons between the processes.

TABLE 4.4

Intermediates obtained by sulphonation

Product	*Starting material*	*Figure*	*Sulphonation process*
Benzene-1,3-disulphonic acid	Benzene	3.2	30% oleum/180°C
p-Toluenesulphonic acid	Toluene	3.6	Monohydrate(100°C
Naphthalene-1-sulphonic acid	Naphthalene	3.7 4.6	98% H_2SO_4/40°C (96% yield)
Naphthalene-2-sulphonic acid	Naphthalene	3.7 4.6	96% H_2SO_4/165°C (85% yield)
Nitrobenzene-3-sulphonic acid	Nitrobenzene	4.1	65% oleum/80°C
1-Chloro-2-nitrobenzene-4-sulphonic acid	OCNB	4.2	
1-Chloro-4-nitrobenzene-2-sulphonic acid	PCNB	4.3	65% oleum/100°C
2-Nitrotoluene-4-sulphonic acid	*o*-Nitrotoluene	4.4	20% oleum/80°C
4-Nitrotoluene-2-sulphonic acid	*p*-Nitrotoluene	4.4	20% oleum/60°C
2-Naphthol-6-sulphonic acid (Schaeffer)	β-Naphthol	4.5	98% H_2SO_4/100°C and separation by fractional salting
2-Naphthol-8-sulphonic acid (crocein)	β-Naphthol	4.5	
2-Naphthol-1-sulphonic acid	β-Naphthol	4.5	SO_3 or $ClSO_3H$/ solvent

Table 4.4

Product	*Starting material*	*Figure*	*Sulphonation process*
4-Nitroanisole-2-sulphonic acid	4-Nitroanisole	4.3	
Naphthalene-1,5-disulphonic acid	Naphthalene-α-sulphonic acid (or naphthalene)	4.6	Oleum/40°C
Naphthalene-1,3,5-trisulphonic acid	Naphthalene-1,5-disulphonic acid	4.6	
Naphthalene-2,7-disulphonic acid	Naphthalene-β-sulphonic acid (or naphthalene)	4.6	Monohydrate/165°C (65% yield +25% 2,6)
Naphthalene-1,3,6-trisulphonic acid	Naphthalene-2,7-disulphonic acid	4.6	20% oleum/180°C through process
2-Naphthol-6,8 disulphonic acid (G)	2-Naphthol-6-sulphonic acid (through process)	4.5	98% H_2SO_4/120°C and separation by fractional salting
2-Naphthol-3,6-disulphonic acid (R)	2-Naphthol-6-sulphonic acid	4.5	
2-Naphthol-3,6,8-trisulphonic acid	2-Naphthol-6,8-disulphonic acid	4.5	40% oleum/120°C through process
Sulphanilic acid	Aniline	4.1	Dry bake/200°C
4-Chloroaniline-2-sulphonic acid	*p*-Chloroaniline		Dry bake/200°C
2,5-Dichloroaniline-4-sulphonic acid	2,5-Dichloro-aniline		65% oleum/80°C
2-Toluidine-5-sulphonic acid	*o*-Toluidine	4.4	Dry bake/220°C
4-Toluidine-3-sulphonic acid	*p*-Toluidine		Dry bake/190°C
4-Chloro-*o*-toluidine-5-sulphonic acid	4-Chloro-*o*-toluidine	4.4	Dry bake
2-Amino-5-methoxybenzene-sulphonic acid	*p*-Anisidine	4.3	Solvent bake ODCB/175°C
2-Amino-5-sulpho-benzoic acid	Anthranilic acid		
Aniline-2,5-disulphonic acid	Metanilic acid	4.1	65% oleum/160°C
Aniline-2,4-disulphonic acid	Sulphanilic acid	4.1	20%oleum/130°/lime-out
m-Phenylenediamine-4-sulphonic acid	*m*-Phenylenediamine		
p-Phenylene diamine-2,5-disulphonic acid	*p*-Phenylenediamine sulphonic acid	4.3	
Naphthionic acid	α-Naphthylamine	4.6	Dry bake/170°C
1-Naphthylamine-4,8-disulphonic acid (S)	1-Naphthylamine-8-sulphonic acid	4.6	10% oleum/100°C
Naphthasultam-2,4-disulphonic acid	S-Acid	4.6	65% oleum/80°C
2-Naphthylamine-1,5-disulphonic acid	2-Naphthylamine-1-sulphonic acid (Tobias)	4.5	20% oleum/25°C

Table 4.4

Product	*Starting material*	*Figure*	*Sulphonation process*
2-Naphthylamine-1,5,7-trisulphonic acid	2-Naphthylamine-1-sulphonic acid (Tobias)	4.5	65% oleum/80°C
Anthraquinone-1-sulphonic acid	Anthraquinone	4.7	20% oleum/HgO catalyst/150°C (75% yield +22% anthraquinone for recycle)
Anthraquinone-2-sulphonic acid	Anthraquinone	4.7	20% oleum/150°C (90% yield after 50% conversion and anthraquinone for recycle)
Anthraquinone-1,5-disulphonic acid	Anthraquinone	4.7	65% oleum/HgO catalyst/125°C (after separation 40% yield of 1,5 +25% 1,8)
Anthraquinone-1,8-disulphonic acid	Anthraquinone	4.7	
Anthraquinone-2,6/2,7-disulphonic acids	Anthraquinone-2-sulphonic acid		40% oleum/150°C
1-Aminoanthraquinone-2-sulphonic acid	1-Amino-anthraquinone	4.7	20% oleum/120°C
p-Toluenesulphonyl chloride	Toluene	3.6	$ClSO_3H$/20°C, isomer separation
m-Nitrobenzene-sulphonyl chloride	Nitrobenzene		$ClSO_3H$/100°C
1-Chloro-2-nitrobenzene-4-sulphonyl chloride	OCNB	4.2	$ClSO_3H$/100°C
Acetanilide-4-sulphonyl chloride	Acetanilide	4.1	$ClSO_3H$/50°C
3-Acetamido-4-methoxybenzene-sulphonyl chloride	*o*-Acetanisidide		$ClSO_3H$/50°C
2,5-Dichloroaniline-4-sulphonyl chloride	2,5-Dichloroaniline-4-sulphonic acid		$ClSO_3H$/50°C
5-(Chlorosulphonyl) salicylic acid	Salicylic acid		$ClSO_3H$/45°C

(a) The substrate is heated with sulphuric acid of specific strength, e.g. concentrated (96% H_2SO_4) or monohydrate (100% H_2SO_4). When reaction is complete the mixture is cooled and drowned-out into water or ice and water. Isolation is normally achieved by filtering off the precipitated product, but where the product is too soluble to give a ready recovery 'lime-out' is necessary. Addition of hydrated lime or chalk neutralises the excess acid with precipitation of calcium sulphate, which can be filtered off to leave a solution of the product for evaporation or use as liquor at the next stage (e.g. nitrobenzene-3-sulphonic acid liquor reduced to metanilic acid).

(b) When more powerful conditions than (a) are required then oleum (a 20% or 65% solution of sulphur trioxide in monohydrate) is added to the sulphuric acid solution of the compound to be sulphonated. (Where the use of these two commercially available strengths of oleum is mentioned in Table 4.4 it should be remembered that this is indicative only of the excess sulphur trioxide at various strengths, depending on the other components in the reaction mixture.)

(c) In certain cases an alternative to the oleum process is to use an organic solvent (e.g. methylene chloride or 1,2-dichloroethane) in place of sulphuric acid, together with sulphur trioxide (stabilised as the trimeric γ-form) or chlorosulphonic acid ($ClSO_3H$). This alternative is environmentally attractive, but the overall cost is often greater than that of conventional sulphonation.

(d) A so-called bake process is economically applicable for the sulphonation of many arylamines. The amine sulphate is pre-formed and heated to about 180°C to rearrange the molecule with introduction of a sulphonic acid group wholly in the *para* position (or the *ortho* position if the *para* is blocked). Bake sulphonation was originally carried out in ovens by 'cooking' the sulphate on trays. This was improved by constructing large rotating metal cylinders or boilers which were directly gas fired. Third-generation 'bakers' have been developed from continuously fed high-temperature dryers. In a few cases dry bake conditions are too rigorous for the arylamine (e.g. anisidines) and the rearrangement is then effected by heating in a high boiling point liquid medium. These reactions are called solvent bakes to distinguish them from dry bakes.

(e) Chlorosulphonic acid is occasionally used for introduction of the sulphonic acid group but more usually the sulphonyl chloride (RSO_2Cl) is the isolated product. Chlorosulphonations can be considered together with sulphonations for convenience.

(f) For the sake of completeness a few important examples of 'sulphition', as an alternative to sulphonation, are included. The introduction of a sulphonic acid by nucleophilic displacement of a reactive chlorine atom is a general reaction exploited in the reaction of 2,4-dinitrochlorobenzene with sodium sulphite to form 2,4-dinitrobenzenesulphonic acid. This can then be reduced to *m*-phenylenediaminesulphonic acid as an alternative process to the sulphonation of *m*-phenylenediamine. A more specific sulphition reaction is used in the manufacture of 1-amino-2-naphthol-4-sulphonic acid (Figure 4.5) where 1-nitroso-2-naphthol is reacted with sodium bisulphite.

4.6 OXIDATION

Although this process (i.e. the introduction of oxygen into or the removal of hydrogen from a molecule) ranks highly in Table 4.1 it mainly occurs as a stage in the primary syntheses, implying that the more highly substituted molecules are less amenable to oxidation. As stated in Chapter 3, the processes for manufacturing anthraquinone, phthalic anhydride, phenol, hydroquinone, benzaldehyde and benzoic acid are virtually all catalytic and sited in dedicated plants. The Hofmann process for making anthranilic acid (Section 3.3.3) should also be mentioned as a special example of hypochlorite oxidation.

If a required intermediate cannot be obtained with the oxidised group functionality already built in at the primary stage, then a batch oxidation is required. Typically this would be used for a substituted benzaldehyde or benzoic acid, with oxidation of the corresponding methyl compound. For example *o*- and *p*-nitrotoluenes (Figure 4.4) are traditionally oxidised to the corresponding nitrobenzoic acids by a mixture of dichromate with sulphuric or nitric acid, but catalytic oxidation processes are available where demand justifies the investment. Other examples include the following:

(a) *p*-Chlorotoluene → *p*-chlorobenzoic acid (oxidising agent: aqueous potassium permanganate)
(b) 2,6-Dichlorotoluene → 2,6-dichlorobenzaldehyde (oxidising agent: manganese dioxide and sulphuric acid)
(c) *O*-(*p*-toluenesulphonyl)-*o*-cresol → salicylaldehyde (oxidising agent: manganese dioxide and sulphuric acid, with hydrolysis)
(d) Toluene-2,4-disulphonic acid → benzaldehyde-2,4-disulphonic acid (oxidising agent: manganese sulphate, potassium permanganate and sulphuric acid).

Two special cases are worthy of mention since they are important intermediates justifying specialised plants. 4,4′-Dinitrostilbene-2,2′-disulphonic acid (Figure 4.4, precursor for the derived diamine and the major class of stilbene FBAs) is manufactured by oxidation of 4-nitrotoluene-2-sulphonic acid in an aqueous alkaline medium by addition of sodium hypochlorite at 80°C, or by using a special catalytic oxidation with air employing a special form of agitation for the aeration [12]. 2-Nitrobenzenesulphonyl chloride (Figure 4.2, precursor for orthanilic acid) is manufactured by the oxidation of 2,2′-dinitrobenzene disulphide by treatment with chlorine in a concentrated acid medium.

4.7 FUSION/HYDROXYLATION

The conversion of arylsulphonic acids to the corresponding hydroxy compound is normally effected by heating with caustic soda (caustic fusion). The primary examples of β-naphthol and resorcinol discussed in Chapter 3 typify the process conditions required. The sulphonic acid and strong caustic liquor are mixed and heated for several hours in the range 250–350°C, allowing water evaporation to take place within limits of the reaction mixture fluidity. A powerful scraper agitator within the cast iron pot is essential. The resultant 'melt' is drowned into water and the product separated after partial or full neutralisation. The co-produced bisulphite gives rise to a sulphur dioxide emission problem during the neutralisation, although absorption in caustic soda produces saleable sodium bisulphite liquor.

In the benzene series, further examples (Figure 4.1) are *m*-aminophenol from metanilic acid and *N,N*-diethyl-*m*-aminophenol from *N,N*-diethylmetanilic acid. In the naphthalene series, where the hydroxyl group is much more commonly introduced into a later stage intermediate, the examples are listed in Table 4.5, including the most important of the letter acids. In most cases selective hydroxylation takes place at the α-position in α,β-disulphonated naphthalenes (e.g. G-acid).

TABLE 4.5

Substituted naphthols obtained by fusion

Product	*Starting material*	*Figure*
1,5-Dihydroxynaphthalene	Naphthalene-1,5-disulphonic acid	4.6
F-acid	Naphthalene-2,7-disulphonic acid	4.6
1,7-Dihydroxynaphthalene-3-sulphonic acid	G-acid	4.5
1-Amino-5-naphthol	Laurent's acid	4.6
J-acid	2-Naphthylamine-5,7-disulphonic acid	4.5
γ-acid	2-Naphthylamine-6,8-disulphonic acid	4.5
H-acid	1-Naphthylamine-3,6,8-trisulphonic acid	4.6
K-acid	1-Naphthylamine-4,6,8-trisulphonic acid	4.6
Chicago acid	Naphthasultam-2,4-disulphonic acid	4.6
Chromotropic acid	1-Naphthol-3,6,8-trisulphonic acid	

In the anthraquinone series conversion of sulphonic acid to hydroxyl groups is not so amenable to caustic fusion and hydroxylation with lime is preferred. The important intermediates anthrarufin (1,5-dihydroxyanthraquinone) and chrysazin (1,8-dihydroxyanthraquinone) (Figure 4.7) are manufactured by heating the disulphonic acid for 20 h at 200°C under pressure with an aqueous suspension of hydrated lime. Alizarin (1,2-dihydroxyanthraquinone) (Figure 4.7) is of lesser importance, but is significant historically, and is a unique example of facile conversion by alkali fusion of a monosulphonic acid (anthraquinone-2-sulphonic acid) to a dihydroxy compound.

A small number of specific intermediates are manufactured by so-called fusion processes where cyclisation rather than hydroxylation is the outcome. Indanthrone (Figure 4.7) is manufactured from 2-aminoanthraquinone by a process involving heating at 200°C with mixed hydroxides of sodium and potassium with elimination of hydrogen. Dibenzanthrone (Figure 4.7) is manufactured from benzanthrone by heating at 150°C with methanolic potassium hydroxide.

A special case of hydroxylation is best exemplified by the manufacture of 1-naphthol-4-sulphonic acid (NW-acid) (Figure 4.6) from naphthionic acid. This may be classed as a 'reverse Bucherer' reaction since hydroxylation is achieved by refluxing with sodium bisulphite. Very rarely can the conversion of amino to hydroxy be achieved by direct caustic fusion, but the alternative manufacturing route to chromotropic acid from Koch acid is one example of this.

4.8 AMINATION

This generic unit process description is used to cover the introduction of amino groups into aromatic molecules by reaction of ammonia or an amine with suitably substituted halogeno, hydroxy or sulphonated derivatives by nucleophilic displacement. Although reaction and operational conditions vary, the process always involves the heating of the appropriate precursor with excess aqueous ammonia or amine under pressure.

For the amination of aryl halides the reaction only proceeds smoothly when the halogen is activated by electron-withdrawing groups. Major examples of suitable precursors are OCNB (Figure 4.2), PCNB (Figure 4.3), 2,4-dinitrochlorobenzene (Figure 4.2), 1,2-dichloro-4-nitrobenzene, 1,4-dichloro-2-nitrobenzene (Table 4.2) for the respective manufacture of ONA, PNA (Section 3.3.1), 2,4-dinitroaniline, 2-chloro-4-nitroaniline and 4-chloro-2-nitroaniline. Manufacture of this family of products calls for an autoclave operating at temperatures of around 175°C and pressures of 3.5–4.0 MPa (500–600 lbf/in^2). In the past explosions under these conditions led IG Farben to develop a generally applicable continuous process based on passage of ammoniacal slurry of the chloro compound through a tube immersed in a heated oil bath [6, 13]. Although this process offers economical advantages in addition to safer operation, these are offset, in multi-product manufacture, by problems in changing over from one product to another. This typifies the assessment of continuous versus batch operation that has frequently to be made by intermediates manufacturers.

Sulphonated nitrochlorobenzenes react under less vigorous conditions and the important 4-nitroaniline-2-sulphonic acid (Figure 4.3) is manufactured in high yield by amination of 4-chloronitrobenzene-3-sulphonic acid with aqueous ammonia at 120°C and 600 kPa (6 atm).

Catalysts (e.g. copper salts) may help with aryl halides of lower activity but amination of chlorobenzene itself is not viable, and amination of 1,4-dichlorobenzene is not established (Section 3.3.1). However the 4-chloro atom of 1,2,4-trichlorobenzene is sufficiently reactive for high-temperature amination to yield, 3,4-dichloroaniline.

Amination of phenolic derivatives is limited to specially developed catalytic processes for aniline (Section 3.3.1) and *m*-toluidine [10]. More general conditions apply to amination of naphthols through a Bucherer reaction in the presence of the bisulphite anion to form the reactive naphthol–bisulphite intermediate (see Figure 4.14).

Important examples of this process are shown in Figure 4.5, with the products Tobias acid, Bronner acid, amido-G-acid and γ-acid. Process conditions of 100–150°C at around 1.5 MPa (200 lbf/in^2) still necessitate lengthy

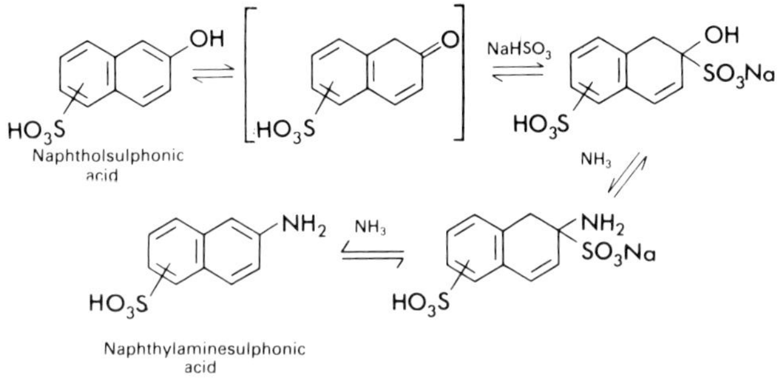

Figure 4.14 – Bucherer reaction for the amination of naphthols

reaction times and the use of batch autoclaves. Excess ammonia is readily flashed off leaving a clean product for filtration.

Where *N*-alkyl or *N*-aryl γ- or J-acids are required it is often preferable to implement the Bucherer reaction between the amine and the corresponding dihydroxy compound, rather than react the amine directly with γ-acid or J-acid [14].

In the anthraquinone series, apart from the special case of the amination of leuco quinizarin (Figure 4.13), sulphonic acid and nitro are the preferred leaving groups. 1-Aminoanthraquinone (Figure 4.7) is manufactured from anthraquinone-1-sulphonic acid or 1-nitroanthraquinone, and similarly betamine (Figure 4.7) from anthraquinone-2-sulphonic acid.

4.9 ALKYLATION

The substitution of a hydrogen atom by an alkyl group can take place at a carbon, nitrogen or oxygen atom to form intermediates. Apart from special cases such as *o*-cresol (Section 3.3.1), *C*-alkylations may be discussed under the headings of alkene reactions and *N*-alkyl rearrangements. In addition to isopropylation of benzene (Figure 3.3) and naphthalene (Figure 3.7), two important examples of alkylation with alkenes are given in Figure 4.15.

Manufacture of *p*-butylaniline, by heating *N*-butylaniline with zinc chloride, typifies the rearrangement reaction appropriate to C_4 and higher alkyl derivatives.

N-Alkylation is most important in producing a wide range of substituted anilines, for use in many dye classes. The products made in the highest quantities, i.e. mono- and di-*N*-methyl- and *N*-ethyl-anilines, are manufactured by continuous catalytic processes (Section 3.3.1), but other products require a wide variety of batch processes. These are summarised with examples below:

(a) Alkyl halides (including benzyl chloride), e.g.
 (i) Metanilic acid → *N,N*-diethylmetanilate, with ethyl chloride (Figure 4.2)

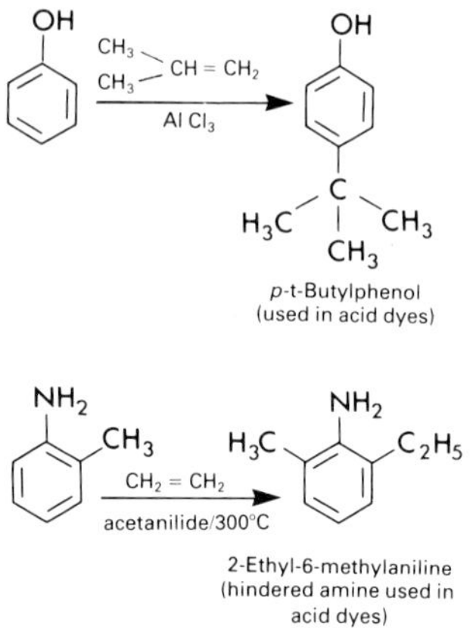

Figure 4.15 – C-Alkylations with alkenes

(ii) *m*-Toluidine → *N*-ethyl-*m*-toluidine and *N,N*-diethyl-*m*-toluidine, with ethanol and hydrochloric acid at 180°C (Figure 4.4)
(iii) *N*-Ethylaniline → *N*-benzyl-*N*-ethylaniline, with benzyl chloride
(b) Ethylene oxide (ETOX), e.g. *m*-toluidine → *N,N*-di(β-hydroxyethyl)-*m*-toluidine, with ETOX in an autoclave at 100°C (Figure 4.4)
(c) Acrylonitrile, e.g. *N*-methylaniline → *N*-(β-cyanoethyl)-*N*-methylaniline, with acrylonitrite
(d) Acrylic acid or ester, e.g. aniline → *N,N*-di(β-methoxycarbonylethyl) aniline with methyl acrylate.

N-Phenylglycine, the key intermediate for indigo, may be manufactured by alkylation of aniline with chloroacetic acid, but it is much more economical, even though three *in situ* stages are required, to use formaldehyde as the alkylating agent, as shown in Figure 4.16.

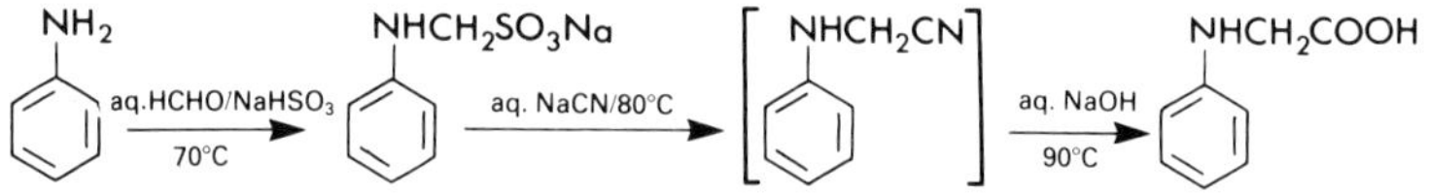

Figure 4.16 – N-Phenylglycine process

N-Alkyl (and *N*-aryl) substituted naphthylamines are also important (e.g. letter acid derivatives) but are usually manufactured by the Bucherer reaction (Section 4.8). Phenyl-peri acid and analogues are unusual in that the phenylation reaction is effected by heating with aniline and aniline hydrochloride in an autoclave.

O-Alkylation is comparable to *N*-alkylation but since the sodium salts are water soluble it is most convenient to react the phenol or naphthol in aqueous caustic solution with dimethyl sulphate or diethyl sulphate. These are comparatively expensive reagents, and functional alkoxy groups are introduced at a prior stage by alkoxylation (Section 4.13) whenever possible. The manufacturing process for cresidine (Figure 4.17) provides a good example of *O*-alkylation where the use of methyl chloride under pressure is preferred to that of DMS at atmospheric pressure.

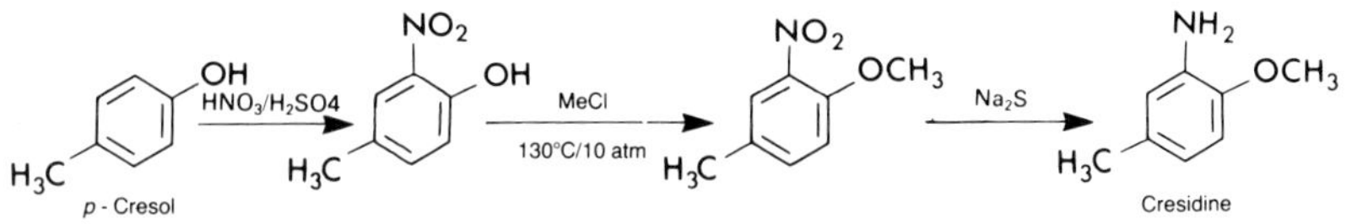

Figure 4.17 – Cresidine process

1,4-Dialkoxybenzenes are manufactured by alkylation of hydroquinone in aqueous alkali, with the choice of agent being determined by economics and plant availability.

Manufacture of alkylsulphones, important intermediates for metal-complex dyes and for reactive dyes, also depends on *O*-alkylation. An arylsulphinic acid (Section 4.4.6) in an aqueous alkaline medium is treated with an alkylating agent (e.g. alkyl halide or sulphate) by a procedure similar to that used for phenols. In the special case of β-hydroxyethylsulphones (precursors to vinylsulphone reactive dyes) the alkylating agent is ethylene oxide or ethylene chlorohydrin.

4.10 HALOGENATION

Halogenation is the process of introducing one or more halogen atoms (F, Cl or Br) into an organic molecule. It is usual to consider ring and side-chain substitution but, for the sake of completeness, hydroxyl substitution to form, for example, acyl chlorides or halogeno heterocyclics is also included here. Chlorine, as by far the cheapest halogen, predominates, but there are advantages of incorporating fluorine or bromine in the end-product which sometimes makes their use economically justifiable [15, 16]. Additionally, brominated intermediates are sometimes necessary, especially in the anthraquinone field, to provide the necessary degree of nucleophilic reactivity.

For chloro-substituted benzenoid derivatives the halogen is introduced at the raw material stage whenever possible through use of chlorobenzene, *o*- and *p*-dichlorobenzenes, chlorotoluenes or chlorophenols (Figure 3.2). Elemental chlorination at a prior stage is the next choice, as with *m*-chloronitrobenzene from nitrobenzene [16], *m*-dichlorobenzene from *m*-dinitrobenzene [8], 3,4-dichloronitrobenzene from PCNB (Figure 4.3), 4-chloro-2-nitrotoluene from ONT (Figure 4.4), 5-chloro-*o*-toluidine from *N*-acetyl-*o*-toluidine (Figure 4.4) and 1,5-dichloroanthraquinone from 1,5-dinitroanthraquinone (Figure 4.7).

When chlorination is necessary at a later stage, neat reactant, as mentioned in connection with the earlier examples, is not so practicable and special conditions are required such as aqueous hydrochloric acid and an oxidising agent (e.g. sodium chlorate). Such conditions are used, for example, in the manufacture of 2,6-dichloro-4-nitroaniline (from PNA or 2-chloro-4-nitroaniline) and 1-chloroanthraquinone (from anthraquinone-1-sulphonic acid (Figure 4.7)). Bromination may be similarly achieved, as in the manufacture of 6-bromo-2,4-dinitroaniline (Figure 4.2).

Side-chain chlorination is virtually limited to the series derived from toluene, of which benzotrichloride has previously been mentioned (Section 3.3.2) as a primary, with benzyl chloride finding significant usage as an alkylating agent. Trifluoromethyl groupings are much more stable than the corresponding trichloromethyl analogues and find occasional usage in low-tonnage intermediates for speciality dye products. This main class of fluorinated intermediates is manufactured by first forming the corresponding trichloromethyl compound and treating this with hydrofluoric acid under pressure to achieve the halogen-exchange reaction [8, 17]. Typical examples of this are benzotrifluoride itself (leading to *m*-aminobenzotrifluoride after nitration and reduction), and *m*-chlorobenzotrifluoride (leading to 4-chloro-2-trifluoromethylaniline after nitration and reduction) from *m*-chlorotoluene.

Introduction of chlorine, bromine and pseudo-halogens (e.g. cyanide) may be achieved indirectly by the application of the Sandmeyer reaction to arylamines. This comparatively expensive special process is reserved for manufacture of particular configurations where the original amine is more readily introduced than a halogen. The arylamine is diazotised normally and then added to the cuprous salt of the appropriate anion. Technically important examples are:

(a) *m*-Chlorotoluene from *m*-toluidine
(b) 3,4,5-Trichloroaniline (after reduction) from 2,6-dichloro-4-nitroaniline
(c) *o*-Chlorobenzonitrile from *o*-chloraniline
(d) 1-Cyanonaphthalene-8-sulphonic acid (for naphthostyril) from peri acid

Direct bromination is chiefly applicable to anthraquinone intermediates since the bromine atom introduced is readily replaced by auxochromic amino or oxy groupings [16]. The most important examples of this are 3-bromobenzanthrone (bromine in nitrobenzene medium), bromamine acid (1-amino-4-bromoanthraquinone-2-sulphonic acid (Figure 4.7) manufactured from 1-aminoanthraquinone-2-sulphonic acid by addition of bromine in aqueous hydrochloric acid), and 1-amino-2,4-dibromoanthraquinone ((Figure 4.7) bromination of 1-aminoanthraquinone suspended in aqueous sulphuric acid).

The more expensive reagents, such as phosphorus trichloride, phosphorus oxychloride and thionyl chloride are used to manufacture speciality acid chlorides or chloro heterocyclics from the corresponding hydroxy compounds. Reaction conditions vary but typical examples are:

(a) *p*-Nitrobenzoic acid → *p*-nitrobenzoyl chloride, with thionyl chloride in toluene solution under reflex (the product is used in toluene solution)
(b) Barbituric acid → 2,4,6-trichloropyrimidine, with phosphorus oxychloride and diethylaniline.

4.11 HYDROLYSIS

The general process definition for hydrolysis embraces all double-decomposition reactions between water (usually in the form of acid or alkali solutions of a wide range of strengths) and an organic molecule. Already considered within this definition have been fusion/hydroxylation reactions (Section 4.7) and specific reactions such as the acid hydrolysis of α-naphthylamine to α-naphthol (Section 3.3.3).

The main group of reactions still outstanding is the hydrolysis of halogenoaryl compounds to hydroxyaryl compounds where milder conditions than encountered in alkali fusion will suffice. Two sets of examples are the aqueous caustic hydrolysis of (a) *o*- and *p*-chloronitrobenzene derivatives to nitrophenols (Figures 4.2 and 4.3) and (b) brominated anthraquinone derivatives (Figure 4.7) to hydroxyanthraquinones such as 1-amino-2-bromo-4-hydroxyanthraquinone (bromo pink).

There should also be mentioned the hydrolysis of *N*-acyl derivatives (Section 4.16) back to the parent arylamine, where the acyl group is frequently used to protect the amine.

4.12 CONDENSATION

This term covers all processes, not previously included in other process definitions, where water or hydrogen chloride is eliminated in a reaction involving the combination of two or more molecules. Enamel-lined reaction units are prevalent in this area because of their versatility under acid (especially hydrochloric), alkali, oxidising or reducing operating conditions. The main process groups collected under this mantle are:
(a) Nitrogen and sulphur heterocycle formation
(b) Amide formation from acid chlorides
(c) Substituted diphenylamines
(d) Miscellaneous cyclisations.

There are an increasing number of heterocyclic intermediates, especially novel diazo [18] and coupling [19] components for azo dyes. Table 4.6 summarises the process conditions for a selection of examples typifying the major ring systems. The benzoxazole example is only one of many heterocyclics specifically for use in the manufacture of fluorescent brighteners; others include triazoles, pyrazolines and coumarins [1].

TABLE 4.6

Typical processes for heterocyclic intermediates

Product	*Process*	*Use*
1-(4′-Sulphophenyl)-5-pyrazolone-3-carboxylic acid	Phenylhydrazine-4-sulphonic acid + oxalacetic ester (Figure 4.1) in aqueous alkaline medium at 60°C	Coupling component for yellow azo dyes including tartrazine
3-Cyano-6-hydroxy-4-methyl-*N*-methylpyrid-2-one	Cyanoacetamide + acetoacetic ester (KOH/MeOH at 70°C)	Coupling components (with variety of *N*-substituents) for greenish-yellow azo dyes
1,2-Bis (5′-methylbenzoxazol-2′-yl)-ethane	2-Amino-4-methylphenol + succinic acid	Dehydrogenated to ethylene derivative as FBA intermediate
2-Aminothiazole	Chloroacetaldehyde + thiourea	Nitrate salt rearranged to 5-nitro-2-amino-thiazole as diazo component for blue disperse dyes
2-Amino-6-methoxy-benzothiazole	*p*-Anisidine + + thiocyanogen (acetic acid + + NaSCN + Br_2)	Diazo component for disperse azo reds
2-(4′-Aminophenyl)-6-methylbenzothiazole	*p*-Toluidine + sulphur for 20 h at 230°C, separated from unwanted primuline base (trimer) by vacuum distillation	Sulphonated derivatives as diazo components for acid yellow azo dyes
3-Amino-1,2,4-triazole	Aminoguanidine + formic acid	Diazo component for cationic azo reds (after methylation)
2-Amino-3-carbomethoxy-thiophene	Mercaptoacetaldehyde + cyanoacetic ester	Nitrated to 5-nitro derivative as diazo component for disperse azo blues [20]
Quinaldine	Aniline + croton-aldehyde (Figure 4.1) (Skraup reaction)	Cyanine dyes and quinophthalones

Table 4.6

Product	*Process*	*Use*
2,3,3-Trimethylindolenine	Phenylhydrazine + methylisopropylketone (Fischer indole synthesis)	Methylated with dimethyl sulphate to give Fischer's base

A major group of carbonamides are the Naphtol products formed by condensation of 2,3-hydroxynaphthoic acid with a wide range of arylamines. The simplest example, Naphtol AS from aniline (Figure 4.5), is typical, and manufacture is accomplished by suspending the acid in a solvent such as toluene, pre-forming the acid chloride by addition of phosphorus trichloride and then adding the aniline.

Sulphonamides, as a class, are simple to manufacture once the isolation conditions for the moderately stable sulphonyl chloride (Table 4.4) have been established. Basically all processes involve the addition of the sulphonyl chloride paste to an aqueous solution of excess ammonia or amine. The product can usually be filtered off in a reasonably pure form with only the hydrolysis product remaining in the liquor.

The reaction of substituted chloronitrobenzenes with arylamines to form substituted diphenylamines is typified by 4-nitrodiphenylamine-2-sulphonic acid (Figure 4.3) where 4-chloronitrobenzene-3-sulphonic acid (PN salt) is condensed with aniline in an aqueous medium at 120°C and 200 kPa (2 atm) in the presence of alkaline buffer with a pH as low as possible to avoid the competing hydrolysis of the PN salt.

Specific cyclisations classed as condensation reactions include the manufacture of benzanthrone (Figure 4.7) from anthraquinone by heating at 150°C with glycerol and iron powder in sulphuric acid.

4.13 ALKOXYLATION

The nucleophilic replacement of an aromatic halogen atom by an alkoxy group is an important process, especially for production of methoxy-containing intermediates. Alkoxylation is preferred to alkylation of the phenol wherever possible and typically involves the interaction of an activated chloro compound (e.g. OCNB, PCNB or DNCB) with the appropriate alcohol in the presence of alkali. Careful control of alkali concentration and temperature are essential and formation of by-product azoxy compounds is avoided by passing air through the reaction mixture [7]. For example, *o*-nitroanisole (precursor to *o*-anisidine and dianisidine (Figure 4.2)) is manufactured by adding methanolic caustic soda solution to a methanol solution of OCNB at 70°C then gradually heating to 95°C to complete the reaction. After diluting with water the product is separated in 90% yield and the methanol recovered from the aqueous layer.

Methoxylation of 4-chloro-3-nitrotoluene (Table 4.2) by an analogous process provides the cresidine precursor 4-methoxy-3-nitrotoluene as an alternative route to alkylation (Figure 4.8).

In the anthraquinone series related reactions are carried out with nitro leaving groups (e.g. 1,5-dimethoxyanthraquinone (Figure 4.7)) or sulphonic acid leaving groups (e.g. 1,5-diphenoxyanthraquinone from anthraquinone-1,5-disulphonic acid).

4.14 ESTERIFICATION

This self-explanatory process is infrequently used since carboxylic esters are comparatively rare substituents. Examples include methylanthranilate, 2-aminodimethylterephthalate and 5-aminodimethylphthalate by processes based on classical esterification using methanol and sulphuric acid. The most important esters are dealt with in Section 4.16 where the 'alcohol' is frequently a substituted *N*-hydroxyalkyl derivative.

4.15 CARBOXYLATION

This is the process of introducing a carboxylic acid group into a phenol or naphthol ring by reaction with carbon dioxide under appropriate conditions of heat and pressure. Apart from the already covered salicylic acid (Section 3.3.1), carboxylation of β-naphthol is the prime example.

A 'dry melt' of sodium β-naphtholate is heated at 240–250°C with carbon dioxide under pressure (600 kPa (6 atm)). This poses equipment problems and the traditional use of cumbersome pressure bakers or special vertical autoclaves is slowly being replaced by more sophisticated equipment or by solvent processes. The conversion to 2-hydroxy-3-naphthoic acid is only about 40% but with recycle of the recovered β-naphthol the overall yield is 75%.

Migration of the inserted carboxylic acid group is possible and this factor is controlled in the manufacture of 2-hydroxy-1-naphthoic acid by limiting the reaction temperature of β-naphthol carboxylation to 140°C. Similarly if the phenol carboxylation temperature is raised to 220°C the first formed salicylic acid is isomerised to *p*-hydroxybenzoic acid.

4.16 ACYLATION

Somewhat analogous to alkylation, the substitution of a hydrogen atom by an acyl group can take place at a carbon, nitrogen or oxygen atom.

The most common method of *C*-acylation is the Friedel–Craft reaction between an acyl chloride or diacid anhydride and an aromatic residue. Typically *o*-carboxybenzophenones are formed by heating benzenoid substrates, phthalic anhydride and aluminium chloride. *o*-Benzoylbenzoic acid (Figure 3.8) is the basic example and substitution of benzene by toluene or chlorobenzene, followed by ring closure, leads to 2-methylanthraquinone and 2-chloroanthraquinone respectively. More important is the special case of quinizarin (Figure 4.12), which is manufactured starting from *p*-chlorophenol, rather than the more expensive hydroquinone. This acylation and ring closure can be effected without aluminium chloride by heating phthalic anhydride and *p*-chlorophenol in sulphuric acid at 200°C in the presence of boric acid.

N-Acetyl and *N*-benzoyl derivatives are formed simply by treating the arylamine in an aqueous medium with acetic anhydride or benzoyl chloride. Apart from the occasional requirement of an *N*-acyl group in the final product, the *N*-acetyl group is frequently used in the manufacturing chain to temporarily protect the amine group during nitration (Table 4.2), chlorination (Figure 4.4) or chlorosulphonation (Table 4.4) since it is readily hydrolysed after reaction and drown-out. Protection of *o*-aminophenols in this way is a slight variation since the *N*-acetyl derivative is readily subject to ring closure to form the 2-methylbenzoxazole (Figure 4.2).

Selective acetylation of diamines is utilised in the manufacture of 4-acetylamino-2-aminobenzene sulphonic acid from *m*-phenylenediamine sulphonic acid and 5-acetylamino-2-aminobenzene sulphonic acid from *p*-phenylenediamine sulphonic acid. In these cases use is made of the increased difficulty of acylating the amino group sterically hindered by a bulky *ortho* substituent. The corresponding dichlorotriazinyl derivatives, using cyanuric acid as the acylating agent, are important reactive dye intermediates.

N-Acetoacetarylides ($RNHCOCH_2COCH_3$) are coupling components in azo pigments that are readily manufactured by treating the arylamine with acetoacetic ester or diketene (Figures 4.2–4.4).

O-Acylation is used to modify side-chain properties and typically achieved by heating the *N*-β-hydroxyethylaniline with acetic anhydride to form the *O*-acetoxy derivative (e.g. *N,N*(β-acetoxyethyl)-*m*-toluidine (Figure 4.4)).

4.17 PHOSGENATION

Reaction of phosgene with arylamines to form ureas, and with reactive aryl species to form substituted benzophenones, are special cases of acylation. They are dealt with separately since more specialised plant is required than for other acylations. Phosgenation of arylamines under different conditions to form isocyanates is not dealt with here since isocyanates find little use as dye intermediates.

Urea formation takes place readily with water-soluble arylamines by simply passing phosgene through a slightly alkaline solution. Important examples of this are carbonyl-J-acid from J-acid (Figure 4.5) and 4,4′-diamino-diphenylurea-3,3′-disulphonic acid from *p*-phenylenediamine sulphonic acid (Figure 4.3).

Michler's ketone (4,4′-bis(dimethylamino)benzophenone) is a vital intermediate for triarylmethane dyes and is manufactured by reacting *N,N*-dimethylaniline (DMA) in two-stages with phosgene in the presence of zinc chloride using excess DMA as acid binder. The separation of the product from DMA and the recycling of the latter are key process steps. The ethyl homologue (ethyl ketone) is manufactured similarly.

4.18 DIAZOTISATION AND COUPLING

Although diazotisation and coupling processes occur in the manufacture of intermediates, e.g. hydrazine formation (Section 4.4.3), Sandmeyer reactions (Section 4.10) and azo double bond reduction (Section 4.4.5), they are the essence of azo dye manufacture and will be dealt with in Chapter 5.

4.19 CONCLUSION

Full comprehension of Chapters 3 and 4 should enable the reader to consider any of the many other intermediates not specifically mentioned and work out possible routes and processes. To test this hypothesis the reader is invited to think about starting materials and routes to the intermediates shown in Figure 4.18.

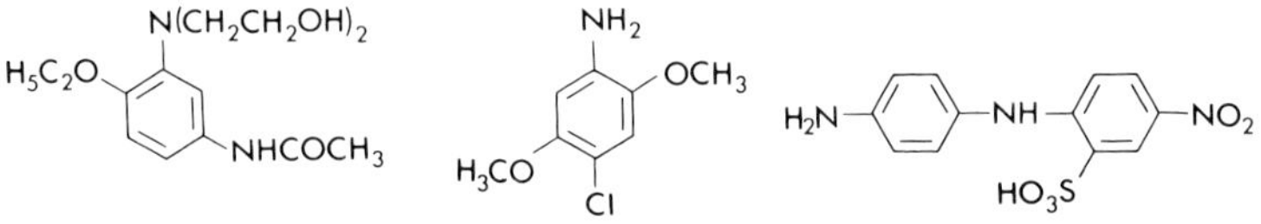

Figure 4.18

REFERENCES

1. Zahradnik, 'Production and application of fluorescent brightening agents' (New York: John Wiley and Sons, 1982).
2. 'Colour Index' 3rd Edn (and supplements) (Bradford: SDC, 1971, etc.).
3. Herrick et al., 'Unit process guide to organic chemical industries' (Ann Arbor Science, 1979).
4. Fierz-David and Blangey, 'Fundamental processes of dye chemistry' (New York: Interscience, 1949).
5. Venkataraman, 'The chemistry of synthetic dyes', Vol. 1 (1952), Vol. 3 (1970) (New York: Academic Press).
6. Groggins, 'Unit processes in organic synthesis' 5th Edn (New York: McGraw Hill, 1958).
7. Donaldson, 'The chemistry and technology of naphthalene compounds' (London: Edward Arnold, 1958).
8. BIOS 986, Part 1.
9. Dittmar, Chem. Rundschau, (Nov 1977) 23.
10. Evans and Frey, ACS Symposium on industrial and laboratory nitrations, **22** (1976) 243.
11. Greenhalgh, Endeavour, **35** (1976) 134.
12. BIOS 1157.

13. BIOS 1147.
14. BIOS 1152.
15. Banks, 'Organofluorine chemicals and their industrial applications' (London: SCI/Ellis Horwood, 1979).
16. Booth, in 'Bromine and its compounds' (London: Benn, 1966).
17. BIOS 986, Part 2.
18. Weaver and Shuttleworth, Dyes and Pigments, **3** (1982) 81.
19. Schwander, Dyes and Pigments, **3** (1982) 133.
20. ICI, BP 1 394 365 (1972).

CHAPTER 5

Colorant manufacture

5.1 CLASSIFICATION

In Table 5.1 sales statistics of dyes in the USA, compiled in 1972 [1], are quoted. This shows the relative size and value of the six most important dye chemical classes. Value figures are included to show again (cf. Table 2.2) the wide variation in the average value of the different classes.

There are such significant differences in the process technologies of the major dye classes that dyestuffs plants tend to specialise in a particular class. Thus multi-product azo or anthraquinone plants are the norm and only rarely does a manufacturing unit make more than one class.

TABLE 5.1

US dye sales in 1972 and *Colour Index* listings

	US sales 1972		
Chemical class	*(lb $\times 10^6$)*	*(\$ $\times 10^6$)*	No. listed in the Colour Index
Azo	124	290	2017
Anthraquinone	48	125	451
Phthalocyanine	17.5	54.5	21
Triarylmethane	8.5	20.5	185
Sulphur (estimates)	15	10	148
Indigoid (estimates)	7	5	121
All others	88	124	~ 1000 (18 classes)
Total	308	629	

Manufacture will be considered under these six headings given in Table 5.1, with only the most important azo class involving specific unit processes (i.e. diazotisation and coupling) not previously discussed in Chapter 4.

It is important to consider, for all classes, the manufacture of the 'chemical' (e.g. colour paste) separately from that of the finished product, even though the two parts are often fully integrated. 'Finishing' (or conditioning) involves changes in physical form, the simplest of which is drying and grinding, and further formulation to convert the chemical as manufactured into the coloured product for sale. The most notable examples where finishing technology is at least as important as production of the colorant itself are in respect of pigments and water-insoluble dyes. However, formulation of water-soluble dyes is becoming an increasingly important parameter in the market.

5.2 AZO DYES AND PIGMENTS

Azo dyes are predominant because they are the most versatile of all the classes. They are not only the largest class but are now expanding the most rapidly, and there are examples in the acid, basic, direct, disperse and reactive dye ranges. They cover every shade from greenish-yellow to black and the chromophore can be used for every fibre, natural and synthetic, as well as for pigments. Product developments over the last 30 years have concentrated in the reactive, disperse and basic fields, with acid dyes for nylon showing a recent upsurge [3]. Product innovation followed by process development, especially with disperse and reactive dyes, leads to specific new plant requirements to complete the cycle.

The 2 000 or so identified azo dyes can be subdivided into: monoazo (836), disazo (834), trisazo and more complex azo (347), although many of these products are obsolescent, and there are a significant number of products that are not registered. The large number of intermediates used in azo dyes are conventionally referred to as diazo, tetrazo and coupling (end) components.

The basic manufacturing process has remained unchanged over the years. The primary arylamine (diazo component) is treated with nitrous acid (usually by the addition of sodium nitrite to an acid solution or suspension) and the derived diazonium compound is reacted (coupled) with a 'coupling component' (usually an aromatic amine or hydroxy compound or a keto compound capable of enolisation). The important dis- and poly-azo compounds are manufactured either by using a diamine or polyfunctional coupling component as starting material or by repeating the process with an aminoazo compound. The coupling reaction is followed by precipitation and filtration. Of course many variations of the basic chemistry theme have evolved. The most important of these is the use of nitrosylsulphuric acid to diazotise weakly basic arylamines such as 2,4-dinitroaniline, 2,6-dichloro-4-nitroaniline and 6-bromo-2,4-dinitroaniline.

More evolution has taken place in plant development, with improvements in materials of construction and operational control having been made. The majority of diazotisation and coupling reactions are carried out in an aqueous medium under mild conditions, and plant requirements tend to be much simpler than those for most other unit processes. Open tanks in series are quite adequate and large wooden vats have been replaced by rubber- or tile-lined mild steel tanks only in recent years because of limitations of size. Figure 5.1 shows diagrammatically the traditional layout, which relies on gravity transfer and simple manual operation.

Even the more complex azo reactive dyes can usually be made in equally simple plant since either a pre-formed reactive diazo or coupling component is used, or the reactive group (e.g. cyanuric chloride followed by ammonia or amine) can be added to the aminoazo base *in situ* within the coupling tank.

The key process parameters of mixing, pH and temperature have all developed independently. Long experience with paddle agitators in vats has given an empirical basis to the much more scientific approach required for the high degree of mixing necessary on scales up to 50 000 l.

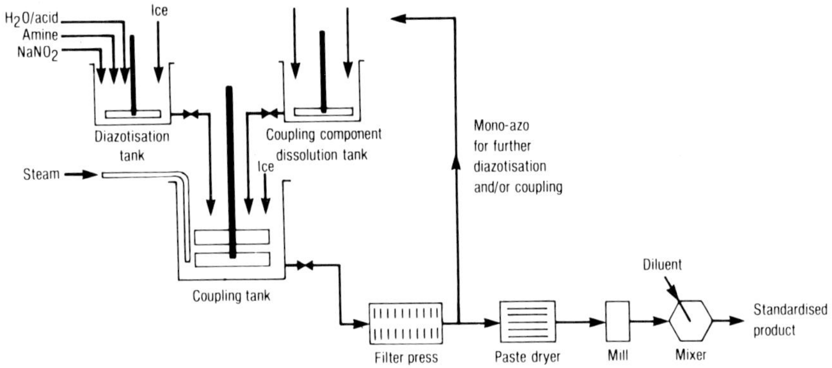

Figure 5.1 – Units involved in azo dye manufacture

The replacement of pH papers by pH meters has developed one step further with automatic pH control, where a fall in pH is programmed to call-up the addition of the required amount of alkali to maintain the process under optimum conditions. The combined importance of pH control and mixing for the diazo coupling reaction has led to much fundamental research of direct technological relevance [4].

Temperature at both diazotisation and coupling stages is traditionally controlled by the use of ice, with live steam heating being employed to meet any higher temperature needed at the isolation stage. Developments have involved integrated ice-making plants (technology from the fisheries industry) interfaced with automatic delivery systems to avoid manual handling. Two published articles give details of the design, layout and engineering of a six-unit plant for 100 different azo dyes (YCL) [5] and a 4 000 tonnes per annum multi-product plant (ACNA) [6]. Reaction units from the former, together with the associated control panel, are shown in Figure 5.2.

A typical plant layout for a modern azo dye plant is shown in Figure 5.3 (BAY) [7]. The main advances from Figure 5.1 are improved materials storage and handling, an integrated ice unit and materials transfer controlled by pumping rather than by gravity. All these developments can be combined in a fully computer-controlled azo batch manufacturing plant (ICI) [8].

Future developments in these diazotisation and coupling unit processes are likely to be based on continuous operation of either or both stages. The

Figure 5.2 – Azo reaction vessels, together with control panel (Yorkshire Chemicals plc)

benefits of continuous coupling are both clear and attractive. Considering the typical batch coupling based on Figure 5.1, there are three variations. Most common is direct coupling, involving the addition of diazo to coupler. Occasionally this addition is reversed and called inverted coupling. The most elegant batch technique, requiring more sophisticated control of addition and pH, is simultaneous coupling where, because of the closer control, diazo decomposition and side reactions are avoided. Where simultaneous coupling is applicable it follows, since a very fast reaction is involved, that continuous coupling, with even further advantages (e.g. reactor size) is possible, provided that matching continuous isolation can be engineered. Further advantages accrue from continuous diazotisation either separately or in-line. In all cases the configuration of the reactor and the efficiency of mixing are critical, and where rapid reaction is achieved cooling becomes much less important.

Although there are many patented processes for both continuous diazotisation and continuous coupling [9–13] there are no published details of the actual exploitation. The best examples, based on the simultaneous coupling concept described above, are with azo pigments and azo disperse dyes, where several companies have facilities for continuous or semi-continuous production. Sandoz has a computer-controlled continuous unit within a

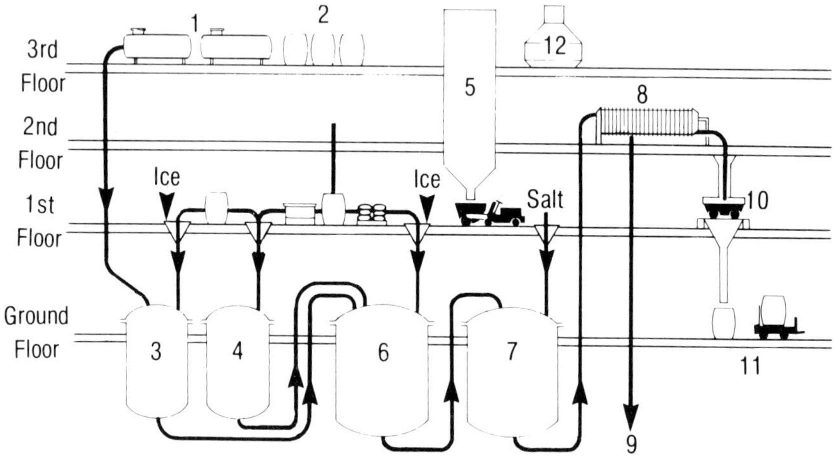

1 – Storage tanks for liquid starting materials
2 – Storage drums for solid starting materials
3 – Diazotisation vessel
4 – Coupling component vessel
5 – Ice machine
6 – Coupling vessel
7 – Isolation vessel
8 – Filter presses
9 – Filtrate to waste liquor treatment plant
10 – Dryers
11 – Emptying of dyestuffs to go to the mill
12 – Outgoing air purification plant

Figure 5.3 – Layout of azo dye manufacturing plant

conventional installation for disperse dyes [14] and ICI Francolor operates a continuous plant for the manufacture of red and yellow azo pigments [15].

5.3 ANTHRAQUINONE DYES

Vat dyes are the main class of anthraquinone-based products. Since they are generally polycyclic products, manufacture tends to involve condensation type reactions at the higher temperature range of conditions, compared with intermediates. A typical example of an important process is shown in Figure 5.4.

Anthraquinone dyes that are members of the acid, disperse and reactive classes are, unlike vat dyes, based on more highly substituted anthraquinone molecules, and thus are made by application of unit processes akin to those used for intermediates. These dyes are complementary to azo dyes, giving brighter colours, particularly reds, blues and turquoises. They are, however, inherently tinctorially weaker than azo dyes and are therefore generally more expensive. New products only appear for highly specialised applications, and in future much more effort is likely to be devoted to expanding azo colorant ranges.

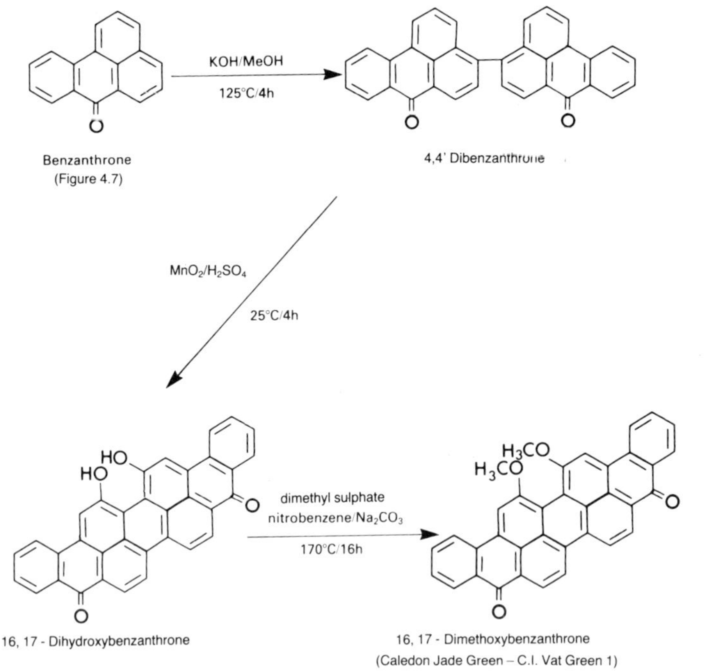

Figure 5.4 – Manufacturing process for C.I. Vat Green 1

Typical process examples are (a) bromamine acid (Section 4.10) aqueous condensations with arylamines to form acid dyes or reactive dye bases (Figure 5.5) and (b) phenoxylation to form a disperse dyes (Figure 5.6). Both of these reflect the more frequent use of solvent processes in anthraquinone dye manufacture, compared with azo and most other classes. A typical anthraquinone dye plant layout (Figure 5.7 [7]) may be compared with the azo dye plant layout of Figure 5.3 to show how the use of organic solvent necessitates additional plant items as well as a solvent-recovery unit. In spite of this increased complexity it was still possible for PCUK (now ICI Francolor) to design a fully computerised vat dye plant [15].

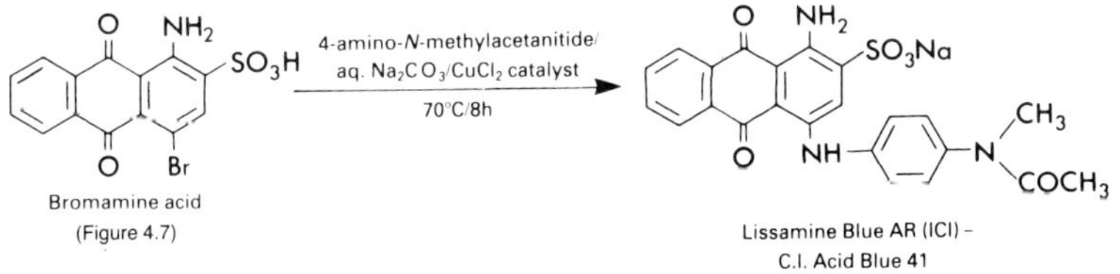

Figure 5.5 – Manufacturing process for C.I. Acid Blue 41

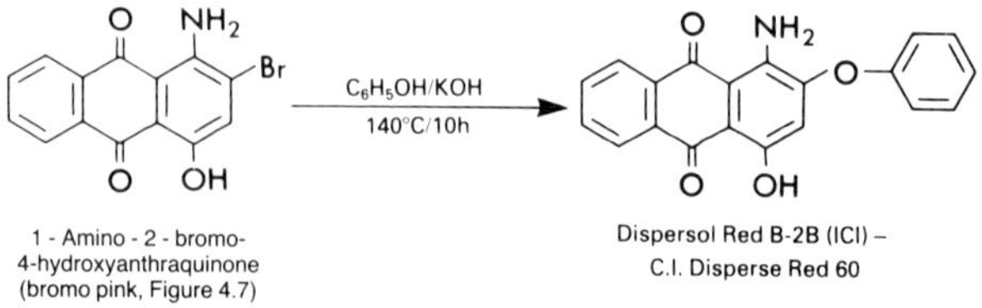

Figure 5.6 – Manufacturing process for C.I. Disperse Red 60

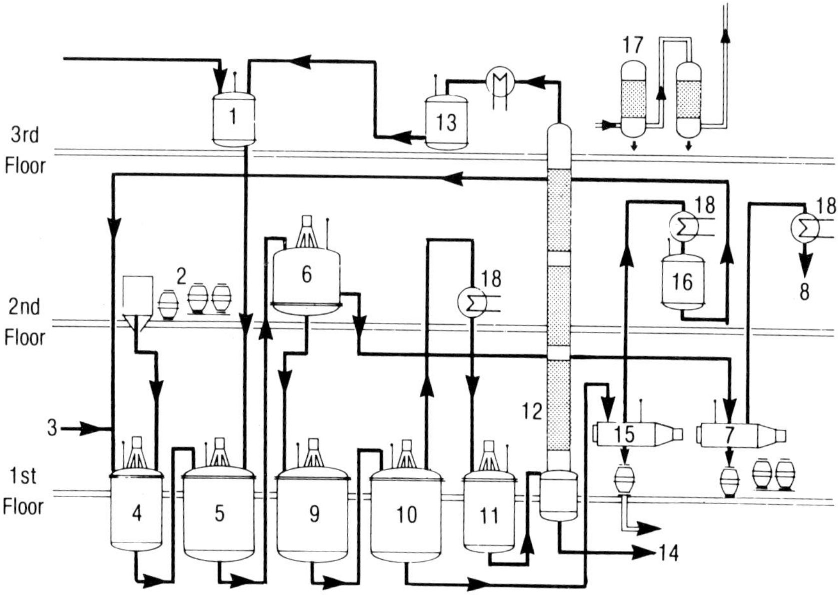

1 – Solvent vessel (coming from storage tank and recovered solvent)
2 – Solid starting materials
3 – Liquid starting materials (including material recovered from the process via 16)
4 – Reaction vessel (fed from 2 and 3)
5 – Dilution vessel
6 – Filter
7 – Dryer for filter cake from 6 (product discharged into drums)
8 – Effluent from dryer to treatment plant
9 – Vessel for filtrate from 6
10 – Distillation vessel for filtrate (stripping)
11 – Receiver for low-boiling fraction from 10
12 – Distillation of low-boiling fraction
13 – Receiver for solvent recovered from 12
14 – Waste products from 11 and 15 go for incineration
15 – Distillation of high-boiling fraction from 10
16 – Receiver for liquid starting material recovered from 15
17 – Outgoing air purification plant
18 – Condensers

Figure 5.7 – Layout of anthraquinone dye manufacturing plant

5.4 PHTHALOCYANINE DYES AND PIGMENTS

Copper phthalocyanine (CuPc) predominates in the phthalocyanine field virtually unchallenged by other metal derivatives. The two original methods of manufacture, i.e. the phthalic anhydride–urea process and the phthalonitrile process, continue to be important [16]. Figure 5.8 summarises how the advantages of the less energy-intensive phthalonitrile process (yield 95%) should be compared with the cheaper raw materials used in the phthalic anhydride process (yield 85%). Plants for both processes are likely to be enamelled cast iron condensation units, to avoid interference from exposed iron, dedicated to the single product to avoid problems of cross-contamination. However, the much shorter reaction time for the phthalonitrile process also opens the way to the possibility for a continuous manufacturing process [17].

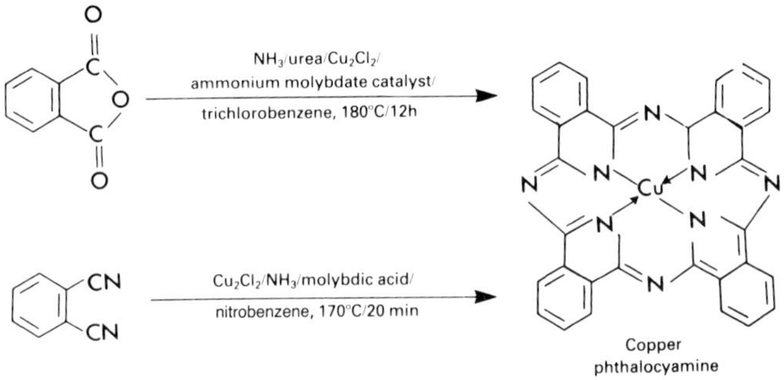

Figure 5.8 – Copper phthalocyanine manufacturing processes

As with azo pigments, it is especially important to differentiate between the chemical and the finished product. Most of the recent developments in copper phthalocyanine technology are concerned with economic production of specific α- and β-polymorphs. The stable β-form is obtained directly from the reaction and requires careful processing to develop its full-strength potential. The α-form, a metastable redder colour, is manufactured from the β-form by reprecipitation from concentrated sulphuric acid (acid pasting) or by 'salt milling' with an inorganic salt that is subsequently washed out. Development and control of these processes provide excellent examples of the need for the application of physical chemistry and X-ray diffraction techniques.

The next important derivatives of copper phthalocyanine are the polychloro ($CuPcCl_{15-16}$) and polychlorobromo (e.g. $CuPcCl_{12.5}Br_3$) green pigments. Dedicated plants are again necessary for the stringent conditions employed and the capital outlay for even a small-scale plant is comparatively large, with the investment risk not justified unless complete technological know-how is available. The most generally applicable process involves adding chlorine or bromine to an aluminium chloride–sodium chloride eutectic mixture at 180–200°C [18]. The only alternative reaction medium for the chlorination is chlorosulphonic acid, which shows little or no advantage over the aluminium chloride melt due to similar handling and corrosion problems. Sulphur dichloride is an alternative chlorinating agent and excess agent may be used as the reaction medium. Reaction conditions are 175°C under pressure, which necessitates an elaborate and expensive plant including a sulphur dichloride recovery system.

There are a large number of water-soluble derivatives of copper phthalocyanine obtained by sulphonation of oleum at 170°C, e.g. $CuPc(3\text{-}SO_3H)_{3-4}$, or chlorosulphonation by heating at 140°C in excess chlorosulphonic acid to give $CuPc(SO_2Cl)_{3-4}$, which is isolated as an unstable paste and treated with amines to give direct dyes such as $CuPc(SO_3H)_{1.5}(SO_2NHC_2H_4OH)_{1.5}$ or turquoise reactive dyes such as $CuPc(SO_3Na)_x(SO_2NHR)_y$, where R is a reactive group and $x + y = 4$.

A comprehensive patent bibliography on phthalocyanine technology has been compiled by Meltzer [19] which summarises the processes for pigments and dyes on a company by company basis. A more recent review [20] summarises developments in both manufacture and finishing, again mainly based on patent information.

5.5 TRIARYLMETHANE DYES

The major products have all been available for almost a hundred years and processes have changed little over the last half century. The simplest way of considering the production of the most commercially important products is under the headings (a) diamino cationic types, (b) triamino cationic types and (c) derived acid dyes.

Malachite green is the simplest, and most important, diamino type. It is manufactured from benzaldehyde and dimethylaniline, and a range of related products can also be made by variation of the alkylaniline and/or use of substituted benzaldehydes.

Although triamino types can be prepared by the aldehyde process, by analogy with malachite green, the required *p*-aminobenzaldehydes are not readily accessible and manufacture is by either a Michler's ketone route or a specific oxidation process.

Acid dyes (usually containing two sulphonic acid groups to give a net negative charge) based on cationic chromophores may be manufactured by a malachite green type process using a benzaldehyde sulphonic acid or *N*-ethyl-*N*-(sulphobenzyl)aniline. Alternatively the activity of the α position in the naphthalene nucleus is exploited by condensation of a benzhydrol with a naphthalenesulphonic acid (e.g. Naphthalene Green V (C.I. Acid Green 16) from 4,4′-bis(diethylamino)benzhydrol and naphthalene-2,7-disulphonic acid). In the triamino series it is more usual to sulphonate the parent triarylmethane dye with oleum.

5.6 SULPHUR DYES

These dyes are complex, with indefinite structures, best considered from the process point of view as 'cookings' of sulphur-containing mixtures to give reasonably consistent products for cheap low-quality dyeings. They are gradually decreasing in importance, in spite of their cheapness, because of increasing environmental constraints on sodium sulphide effluent in both manufacture and use.

The simplest sulphur black (C.I. Sulphur Black 1) is manufactured from 2,4-dinitrochlorobenzene or 2,4-dinitrophenol (either of which may contain impurities or additives such as mononitro compounds) by refluxing with sodium polysulphide liquor for many hours. The dye is precipitated by air oxidation and separated, washed and dried by conventional methods.

An example of another important product obtained from less simple raw materials is Hydron Blue (C.I. Vat Blue 43) manufactured by thionation (with polysulphide in refluxing butanol or cellosolve) of the indophenol derived from the condensation of *p*-nitrosophenol with carbazole. Variations on this product theme allow the process to absorb proportions of organic by-products to minimise the raw material costs.

There is little of general process interest in the cookery of the wide range of thionated products. For further detail the reader is referred to the most recent reviews [21–23], although sulphur dye manufacturers are particularly secretive about process details.

5.7 INDIGOID DYES

With thioindigold dyes, as with triarylmethane and sulphur dyes, there is little new product or process technology to report. These dyes have a traditional and fairly static market, but are tending to be superseded by more sophisticated products.

Only a few years ago similar comments might have been made about indigo itself. However, due to the fashion boom in blue denim, there has been a world shortage of indigo in recent years. A previously contracting market has led to a situation where there are only four companies in the world still manufacturing indigo: BASF, Buffalo Color Corporation (formerly part of Allied Chemicals), ICI and Mitsui, the first three having announced 'debottlenecking' (a word used to denote expansion of capacity by modification to parts of an existing plant rather than constructing a new plant) schemes to permit production to reach 3 000 tonnes per year each.

All the major producers use the long-established route via phenylglycine (I, Figure 5.9) but the supply shortage has prompted several manufacturers to resuscitate alternative routes. The main alternative, via phenylglycine carboxylic acid (II, Figure 5.9), starting from the more expensive anthranilic acid, has a significantly higher materials cost. Its merit is that the individual stages may be made in existing plant whereas the established route would require very large investment in new plant. This makes manufacture of indigo by the existing process a classic example of the need to evaluate cheaper material costs on the one hand against higher capital requirements for any new investment on the other.

For the established process a lower fusion temperature and enhanced yields are possible through use of mixed sodium/potassium salts of phenylglycine together with mixed hydroxides of sodium and potassium. Figure 5.10 shows diagrammatically how the fusion autoclave is fed with the three components and after reaction discharged to an oxidation vessel.

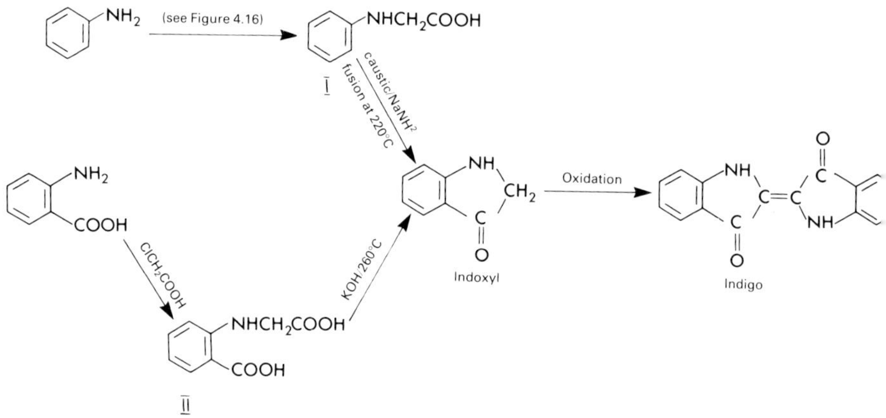

Figure 5.9 – Alternative manufacturing processes for indigo

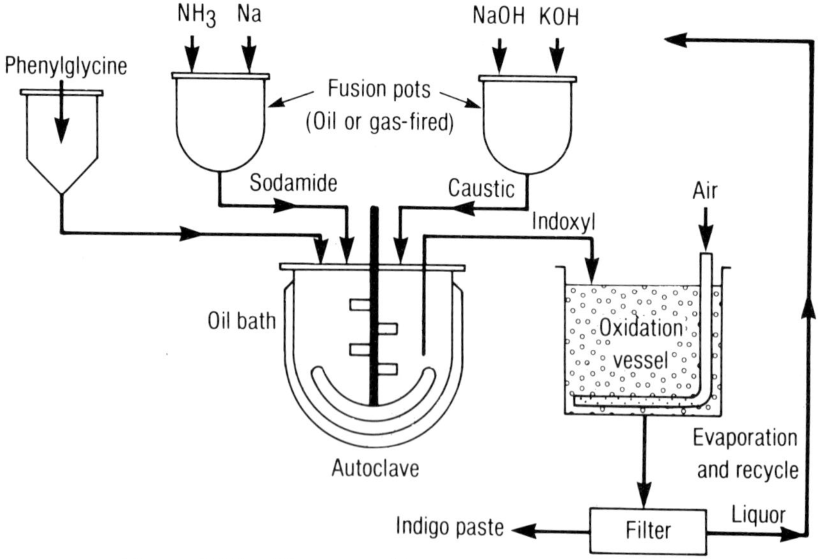

Figure 5.10 – Diagram of indigo manufacturing plant

5.8 OTHER DYE CLASSES

The relatively high figure quoted in Table 5.1 for miscellaneous dyes is somewhat misleading since it includes azoics and FBAs (mainly stilbene) covered, from the manufacturing point of view, as intermediates. Within the other 17 identifiable sub-groups are nitro and methine dyes, both of which are manufactured by intermediates-type unit process technology.

A large miscellany of heterocyclic dyes (e.g. azine, oxazine, quinoline, thiazine, triphendioxazine and xanthene types) have individual process conditions too varied for individual discussion here and which, in any event, justify very little additional commentary to that given in the standard references [24, 25]. The only important addition over recent years to this miscellaneous group is the small number of quinacridone pigments. A manufacturing process for the typical, and most important, red linear *trans*-quinacridone is shown in Figure 5.11. This involves condensation and amination stages, which have no exact parallels in their unit process groups. The process provides a good example of 'telescoping' stages by carrying out the second and third steps consecutively in a single reactor without isolation. Interestingly, by use of a different reaction medium, the first and second steps can be telescoped in an alternative process. As with many other pigments, there is a process requirement that the final product should be obtained in a crystalline modification readily convertible to the useful pigment form.

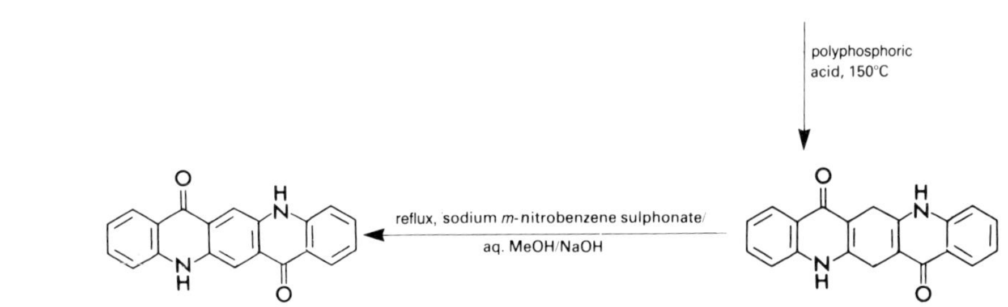

Figure 5.11 – Manufacturing process for quinacridone

5.9 FORMULATION AND FINISHING

As stated in Section 5.1, the finishing processes are often vital to the production process. In the past the distinction was more clear cut since colour paste was usually shipped to a separate 'standardising' department for drying, grinding/milling and formulation to match an agreed standard. The role of the colourist in the manufacturing industry was so important as to require his approval for every single batch of product compared with the standard.

In more recent years standardisation has become more integrated with colorant manufacture. Drying is frequently 'in-line' with colour production (Figures 5.3 and 5.7) and where modern continuous driers have replaced ovens it is often convenient to 'wet standardise' so that final product is produced from the dryer. This is only possible through the use of rapid instrumental matching, with the role of the colourist changing to the establishment and monitoring of agreed testing methods for the standard product, together with definition of acceptable limits of variation.

Although the implementation of finishing may have changed dramatically the principles remain the same. Acceptable diluents together with dispersing agents, wetting agents and/or dedusting agents as required are selected in accordance with shade and strength criteria established at the new product stage. These should be compatible with any variations likely to arise at the final stage of manufacture prior to standardisation. The diluents may be neutral inorganic salts such as common salt or Glauber's salt, but if electrolytes interfere with product properties then more expensive diluents such as urea or dextrine may be utilised. Occasionally dyes are isolated as the insoluble free acid and an alkali salt (sodium carbonate, sodium bicarbonate or trisodium phosphate) is added as a diluent to form the sodium salt on dissolution. Reactive dyes are a special case in requiring addition of a buffer (typically mono- and di-sodium phosphate) to preserve hydrolytic stability on storage.

If shade variation is possible at the manufacturing stage then permissible shading colours require identification and evaluation. To meet market needs it is often necessary to produce a given dye at more than one strength or 'brand'. Water-soluble dyes are being increasingly produced in granular rather than the conventional powder grades. This is effected by the use of more sophisticated drying techniques and meets the need for improved handling and dispensing priorities associated with revision in standards of industrial hygiene [26].

The above principles apply equally to 'liquid brands', whether as solutions of water-soluble products (as with basic dyes) or ultrafine non-settling dispersions (as with disperse dyes). Thus water, or occasionally an organic solvent, becomes the diluent but more adjuvants are usually necessary to counter additional problems associated with liquid brands in respect of pH, viscosity, stability, freezing and evaporation. These liquid brands are usually developed for the benefit of the customer who is not always willing to pay a premium over the powder brand. The manufacturer may therefore have to bear additional costs associated with more demanding formulations, although these may be offset by savings in drying costs if the 'press paste' can be used directly.

Physical form specifications need to be compared with the many other parameters involved in standardisation [27], including health and safety aspects to both manufacturer and user (Chapter 7).

The product has finally to be packed for sale and passed into a transport and distribution system. Important choices are necessary in respect of size and composition of the containers. Drums or smaller packages may be of metal, plastic or fibreboard depending upon technical requirements, coupled with the balance between cost and customer appeal factors. The customer viewpoint in respect of handling and storage has been reviewed by Holme [28].

REFERENCES

1. US International Trade Commission on Production and Sales for Dyes and Pigments (1972).
2. 'Colour Index', 3rd Edn (Bradford: SDC, 1971).
3. Stead, Rev. Prog. Coloration, **6** (1975) 1.
4. Zollinger, Chemtech, (1979) 578.
5. Simborowski, Text. Dyer and Printer, (Jan 1978) 47.
6. Baruffaldi, Impianti, **13** (1980) 509.
7. Boresch, Defazet, **33** (6/7) (1979) 25.
8. Pinkerton and Robinson, Inst. of Electrical Engineers Conf., Publication No. 104 (Oct 1973).
9. BASF, BP 1 082 636 (1963).
10. BAY, German P 2 741 925 (1977).
11. CGY, BP 1 547 759 (1975).
12. HOE, German P 2 635 778 (1975), 2 825 655 (1978).
13. ICI, BP 2 014 597A (1978).
14. Sandoz International Gazette (Sept 1978).
15. ICI, Organics News (July 1982).
16. Booth in 'The chemistry of synthetic dyes', Vol. 5, Ed. Venkataraman (New York: Academic Press, 1971).
17. BASF, BP 1 145 257 (1967).
18. ICI, BP 850 232 (1959), 932 943 (1961).
19. Meltzer, 'Phthalocyanine technology' (NDC, 1970).
20. Moser and Thomas, 'Phthalocyanines', Vol. 2: manufacture and applications (CRC, 1983).
21. Heid, Holoubek and Klein, Melliand Textilber., (1973) 1314.
22. Orton in 'The chemistry of synthetic dyes', Vol. 7, Ed. Venkataraman (New York: Academic Press 1974).
23. Heid in 'Ullmanns Encyklopadie der technischen Chemie', B21 (1982) 65.
24. Venkataraman, 'The chemistry of synthetic dyes', Vol. 2 (New York: Academic Press, 1952).
25. Lubs, 'The chemistry of synthetic dyes and pigments' (New York: Reinhold, 1955).
26. Skelly, J.S.D.C., **96** (1980) 628.
27. Park and Shore, Rev. Prog. Coloration, **12** (1982) 1.
28. Holme, J.S.D.C., **94** (1978) 375.

CHAPTER 6

Unit operations

6.1 INTRODUCTION

The generally accepted definition of unit operations was given by A D Little in 1915: 'Any chemical process on whatever scale conducted may be resolved into a coordinate series of what may be termed unit operations, as pulverising, drying, roasting, crystallising, filtering, evaporating, electrolysing and so on.' Manufacturing plants are composed physically of a series of plant items (e.g. reactor/filter/dryer) performing a series of unit operations in their proper sequence. The capacity of a plant (i.e. how much it can make) is no greater than that of its weakest unit.

There are many ways of categorising these unit operations and some 25 operations have been identified [1]. The most relevant to dyes and intermediates manufacture are as follows:

(a) Materials handling
(b) Absorption and adsorption
(c) Distillation
(d) Liquid–liquid extraction
(e) Solid–liquid separation (filtration)
(f) Evaporation
(g) Crystallisation
(h) Agitation
(i) Drying
(j) Size reduction of solids.

All of these have been mentioned, at least in passing, in the process descriptions of earlier chapters.

Over the last 50 years unit operations have developed more rapidly than has process technology [2]. Part of this evolutionary process has been accompanied by changes in instrumentation and, more recently, computerisation, with a concomitant reduction in manual operations [3]. Improvements in materials handling have also been important in reducing the need for manpower and introducing different patterns of work for operators. Over the last few decades shovelling has been replaced by mechanical handling, and new generations of instrument engineers and computer operators have replaced manual workers.

Analysis of some of the more important unit processes covered in Chapters 4 and 5 shows how the individual unit operations link together, the most generalised sequence being: materials handling – (reaction) agitation – filtration – drying – grinding, shown diagrammatically in Figure 6.1. These unit operations are all well covered in basic chemical engineering textbooks [4, 5] but little has been written that is specifically applicable to the dyestuffs industry. The reason for this is that virtually no operation, or the equipment required, is specific to dyestuffs. Indeed it might be said that most of dyestuffs manufacture has evolved to use equipment initially designed for different (and larger-scale) duties within different industries. However, it is possible to select some of the key unit operations for discussion in terms of dyestuff manufacturing.

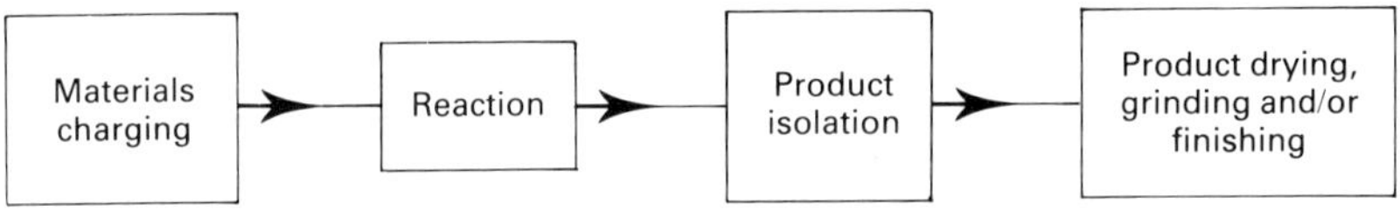

Figure 6.1 – Operation sequence in dye manufacture

The reactor itself is usually the focal point of the plant, but this does not mean that it is the most important manufacturing stage, nor that it absorbs most of the capital or operational costs. Operations subsequent to reaction are often referred to as work-up stages. These vary from product to product with intermediates (used without drying wherever practicable) needing less 'finishing' operations than colorants.

Distillation is often a key stage in isolation and purification of liquid intermediates (e.g. arylamines, Section 4.4.1) or recycling solvent as shown in Figure 5.7. This operation is not as important to the dyestuffs industry as solid–liquid separation involving crystallisation, isolation by filtration and finally drying.

6.2 MATERIALS CHARGING

The options for materials charging at the 'front-end' of the plant are shown diagrammatically in Figure 6.2. The liquid feed vessels or solids-charging equipment are variously connected with the appropriate reaction vessel. In the case of an open pan or vat, alignment of the connections, of which there may be several, is simpler than in the case of a closed vessel, which may already require several 'charge-holes' to be utilised for the measurement of process parameters.

Charging to reactors during operation at greater than atmospheric pressure is virtually limited to gases already handled and stored under pressure, such as hydrogen and ammonia.

The terminology used in Figure 6.2 should be self-explanatory and wide ranges of hoppers, feeders, conveyors, charge-vessels and batchmeters are commercially available. Charging of powders may be subject to electrostatic hazards and charging of flammable liquids also needs special consideration. Controlled addition of pastes is clearly more difficult than is the case with liquids and powders and is not therefore usually so precise.

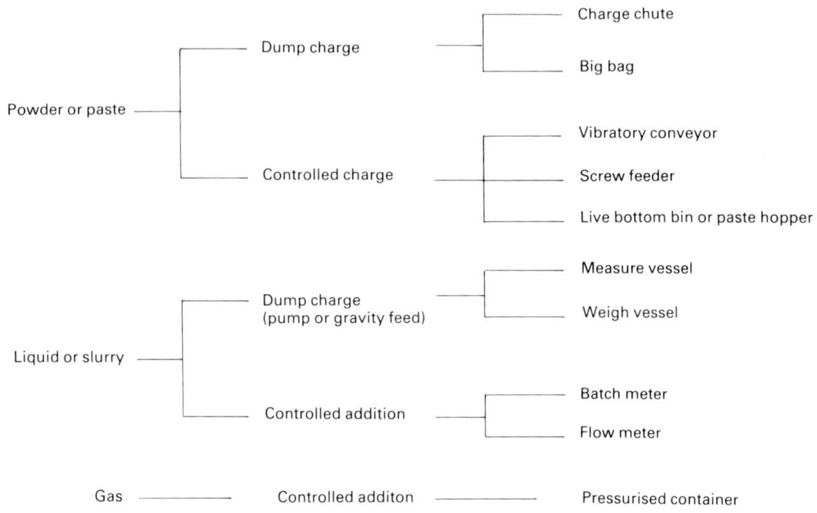

Figure 6.2 – Materials charging

Inspection of the processes described in Chapters 3, 4 and 5 reveals that many reagents have specific handling and charging requirements. For example, nitric acid, oleum and chlorosulphonic acid require specific materials of construction and venting arrangements. Similarly there are special considerations relating to the discharge of gases. For example, sulphur dioxide effluent gas calls for a simple alkali absorption system but other acid gases, such as hydrogen chloride, may form mists that require more sophisticated systems.

The materials handling operations that precede charging, and similarly those involved with isolated product, are important and must be matched up with the plant configuration. Solids, in sacks or drums, are moved about on pallets by fork-lift trucks, newer plants are specifically designed to accommodate this type of materials handling. Liquids are also transported in drums, but, wherever justifiable, in-plant bulk storage together with piped delivery is the preferred option.

Ice is increasingly produced in-plant using modern ice-makers, with decreased manual handling. Where large amounts of ice are required in azo dye plants it can be held in large (50 tonne) bunkers and delivered pneumatically, typically at a rate of 15 tonnes per hour.

6.3 REACTION AND AGITATION

The manufacturing processes described in earlier chapters for intermediates and colorants are, for the reasons given, almost all carried out on a batchwise basis in agitated vessels. These vessels have capacities ranging from 1 000

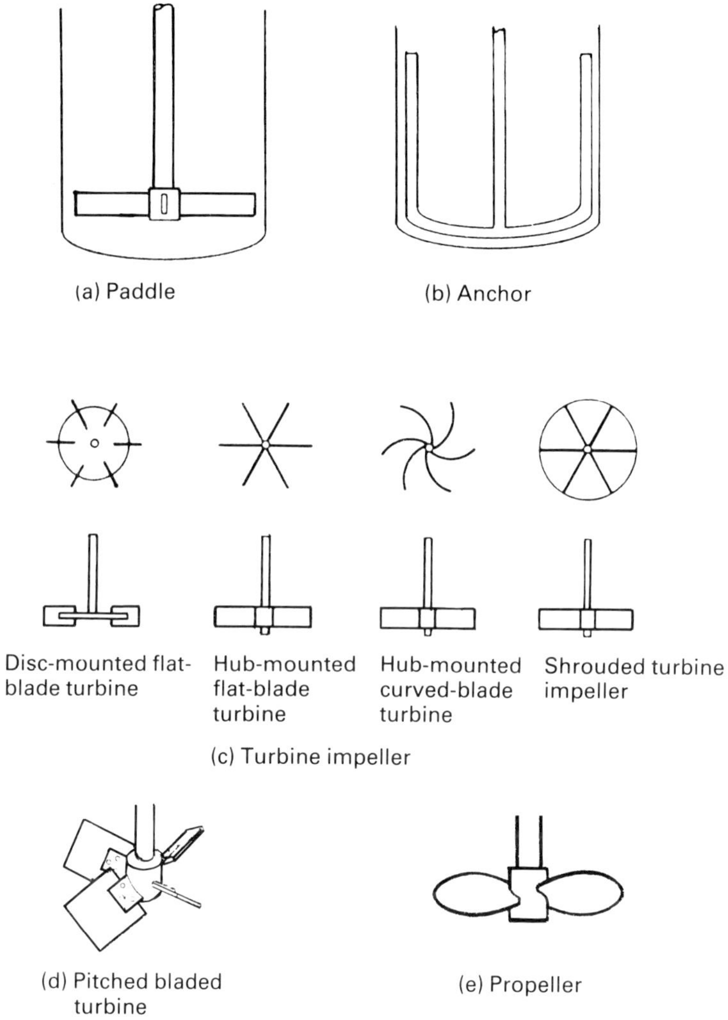

Figure 6.3 – Agitator types

to 45 000 l (250 to 10 000 gallons) and are made from cast iron, stainless steel or other varieties of steel which may be lined with glass (enamel), lead, rubber, bricks or tiles. Very occasionally titanium or Hastelloy reactors may be constructed for special needs. Associated work-up tanks may be made from materials such as reinforced fibre-glass. Wooden vats, and the coopers needed to construct them, have virtually disappeared.

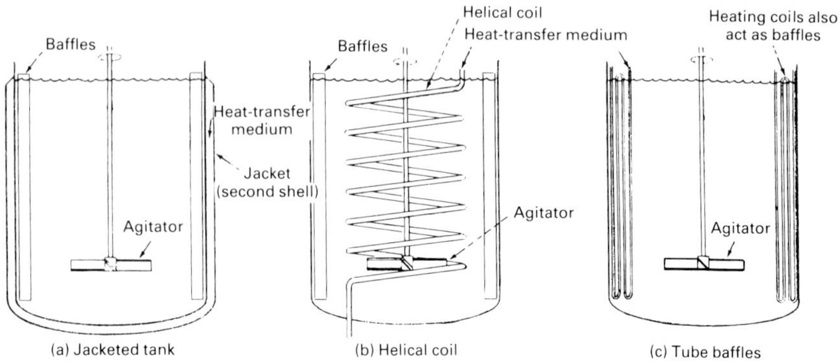

Figure 6.4 – Turbine agitation with baffles and heat transfer

Any reactor, or reaction pan, will be constructed from materials previously tried and tested under manufacturing conditions. It will be fitted with appropriate agitation, heating and/or cooling facilities, and a venting arrangement, which may be a condenser with or without an associated scrubbing system for off-gases. In addition, there will be a variety of measuring systems for monitoring temperature, pressure and pH.

The phrase 'appropriate agitation' is by no means as simple as it sounds since optimisation of the degree of mixing required, viscosity and suspension factors, together with heat-transfer requirements, is a complex design matter. Highly specific agitation systems are only appropriate to single-product units. As multi-product units predominate, the agitation system is usually designed to be a compromise to meet the demands of both the more rigorous and the more tolerant processes.

Both low-speed and high-speed agitators are used. Figure 6.3 shows the main features of the two basic low-speed systems of paddle and anchor, together with the three basic high-speed impeller types. All of these are appropriate to either open vats or closed reaction vessels with a variety of refinements. The simple paddle, for example, can be adapted to use vertical or inclined blades and multi-stage crossed arms.

The geometry of the vessel, coupled with the type of agitation used, dictates any need for baffles. Figure 6.4 shows how three commonly used methods of heat transfer, for heating or cooling, are integrated with a turbine agitator. The use of coils can, however, lead to problems of material hold-up when discharging a thick suspension.

Figure 6.5 shows a typical jacketed reactor (enamel lined) with a cut-away demonstrating alternative agitation arrangements based on anchor or impeller plus baffle agitation arrangements. The configuration of charge hole, temperature probe and venting is fairly standard to most reactors, around a central agitator together with its appropriate gland, bearing, gearbox and power supply. This type of reactor in operation at the 1 000 l scale, for intermediates production, is shown in Figure 6.6.

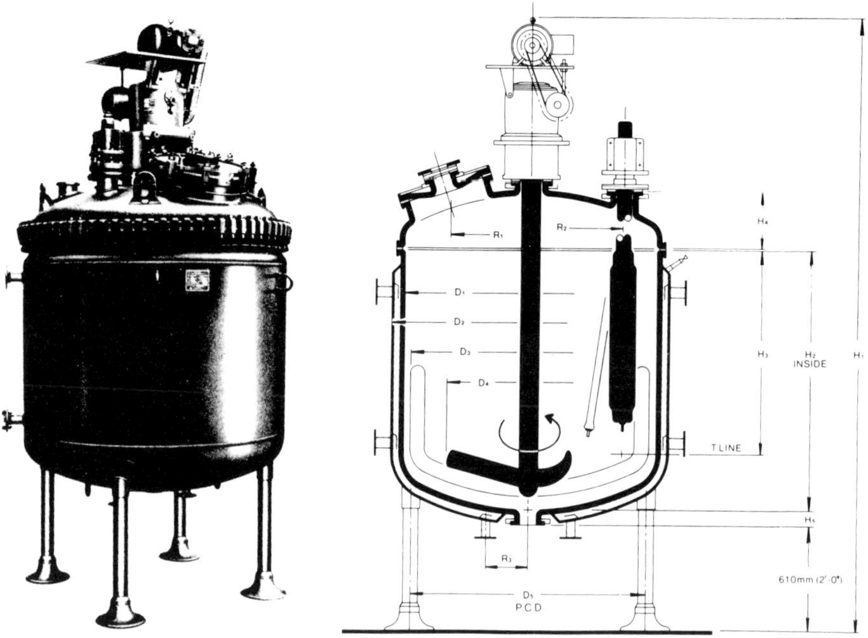

Figure 6.5 – Jacketed reactor vessel (Pfaudler–Balfour)

Figure 6.6 – Jacketed reactor vessel being used in the production of intermediates

6.4 FILTRATION

Before various options available for solid–liquid separation can be considered the physical form of the product needs to be optimised. Crystallisation in the reactor or, as is often the case, in an isolation or drown-out vessel requires careful control to obtain a product with a consistent and reproducible physical form. The correct filter should be chosen to give the best result in terms of speed of filtration and maximum liquid removal. The suspension to be filtered is typically in the range 1–10% solids content, this being as high as possible to allow efficient stirring of the suspension.

Crystallisation is effected by one or a combination of the following techniques:

(a) Variation of reaction conditions
(b) Cooling solution below crystallisation point
(c) Solvent evaporation
(d) Drown-out by dilution with water (usually by addition to water)
(e) Adjustment of pH with acid or alkali to form precipitate
(f) 'Salt-out' by addition of an inorganic cation (usually sodium in the form of common salt).

This is an important aspect of process development that is frequently not given the attention it deserves since, in addition to all the physical chemical aspects, there are many subtleties in both plant configuration and operation. The problems of physical form are frequently magnified when several products are to be isolated using a single filter.

The choice of filter is based on the physical form of the product, while bearing in mind the availability of existing filtration capacity. In practice this means that the physical form may have to be optimised to match an existing filter or, in a new plant, the best combination of physical form with cost-effective filtration needs to be developed. An additional criterion, which is often a complication, is the requirement for washing the product.

The options for product isolation filters will be considered separately under the headings:

(a) Nutsch filters (vacuum filters)
(b) Pressure filters
(c) Filter presses
(d) Diaphragm presses
(e) Continuous vacuum filters
(f) Centrifuges

together with:

(g) Screening filters.

6.4.1 Nutsch filters

This is the simplest type of batch filter and, depending on the product requirement, is constructed to operate under gravity or vacuum. It consists of a tank or box with a perforated base supporting the filter medium; in essence it is scaled-up from the equivalent laboratory device. Figure 6.7 shows the principle of filtration as generally applicable to all filters but having specific applicability to the nutsch.

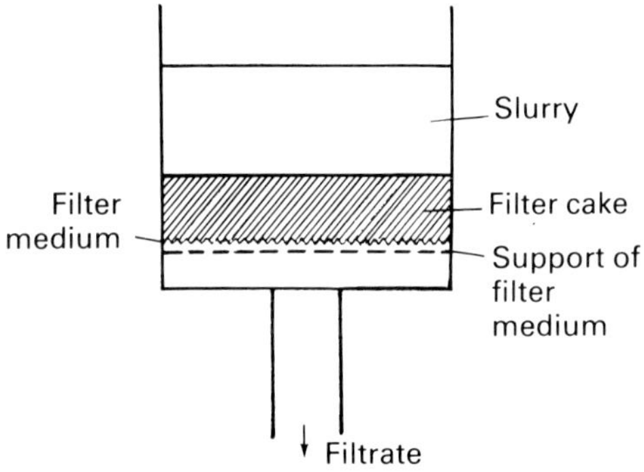

Figure 6.7 – Principle of filtration

The filter medium is either woven fabric (e.g. cotton twill or, in more recent years, synthetic fibre) or ceramic porous tile. The latter is capable of collecting particles down to 1 μm compared with 10 μm for fabric. This means that blinding (clogging of the pores) is potentially more of a problem, although this may be offset by greater ease of cleaning.

In the past many intermediates were filtered on nutsches but these are gradually being replaced by more sophisticated alternatives, which are both more efficient and avoid the danger of exposing workers to potentially harmful substances during the labour-intensive 'digging out' process.

6.4.2 Pressure filters

The pressure filter is essentially an enclosed nutsch operated under pressure from above rather than by vacuum from below. The main advantage, other than enclosure of solvents or other undesirable vapours, is the integrated facility of mechanical discharge using arms on a central shaft. These arms may also be used as an agitator to reslurry the press cake for very efficient washing. Figure 6.8 shows a unit designed by Rosenmund having a central discharge tube which is lowered at the discharging stage of operation. Other designs are operated with side discharge but the principle is the same.

6.4.3 Filter presses

The oldest alternative to the nutsch for batch filtration is the plate and frame press, commonly used for both dyes (Figures 5.1 and 5.3) and intermediates.

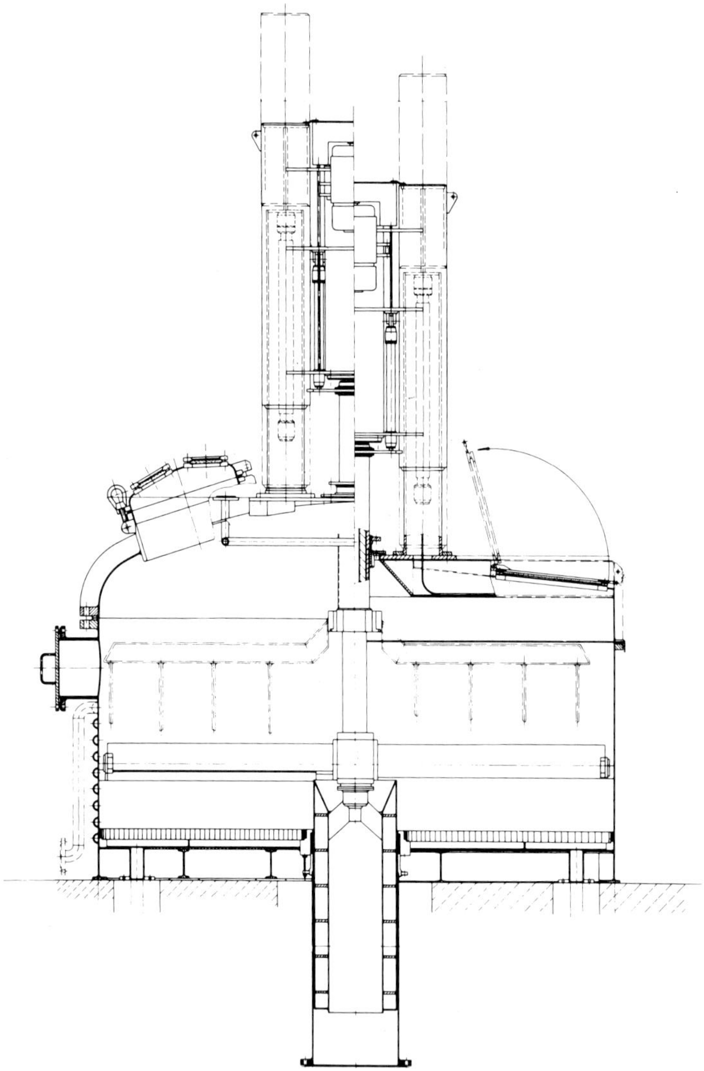

Figure 6.8 – Pressure filter (Rosenmund Ltd)

The filter press is still probably the most flexible multi-product filter because it provides a large filtration and washing area economically.

The simplest form has been developed further through the use of polypropylene or rubber-covered metal plates and frames to replace the original wood furniture. Recessed plates and automatic operation have become standard in newer installations. An external view of a typical small mechanised press is shown in Figure 6.9.

Figure 6.9 – Filter press (Edwards and Jones Ltd)

6.4.4 Diaphragm presses

The recessed plate filter press has been developed further by equipping alternate filter plates with an inflatable membrane. The membrane filter press, or diaphragm press, produces the same residual paste water content at a filtration pressure which is about 50% lower. With filtrations requiring an after-wash, diaphragm presses give higher product paste strengths than conventional filter presses at a much reduced cycle time.

This concept has been further developed by Hoesch into a vertical automatic filter press as pictured in Figure 6.10 (also, somewhat confusingly, called a pressure filter). A different approach to the diaphragm principle is represented by the cylindrical VC filter (Figure 6.11) which has a single large chamber with up to 6 m^2 filtration area. Since this is only a fraction of the filtration area of filter presses, the VC filter requires a larger number of cycles for high paste volumes even though the paste depth may be greater.

Figure 6.10 – Automatic vertical pressure filter (Hoesch)

Figure 6.11 – Variable chamber filter (VC Filters International Ltd)

6.4.5 Continuous vacuum filters

Where automated continuous operation is required to meet process or other needs there are several alternatives based on the principle of a moving, rather than static, filter zone.

The rotary vacuum drum filter (Figure 6.12) has been widely used in continuous single product plants, since a wide range of filter fabrics are available and it has a particularly high wash efficiency. Disc filters work on a similar principle to rotary filters but consist of several thin discs mounted on a shaft in place of the drum.

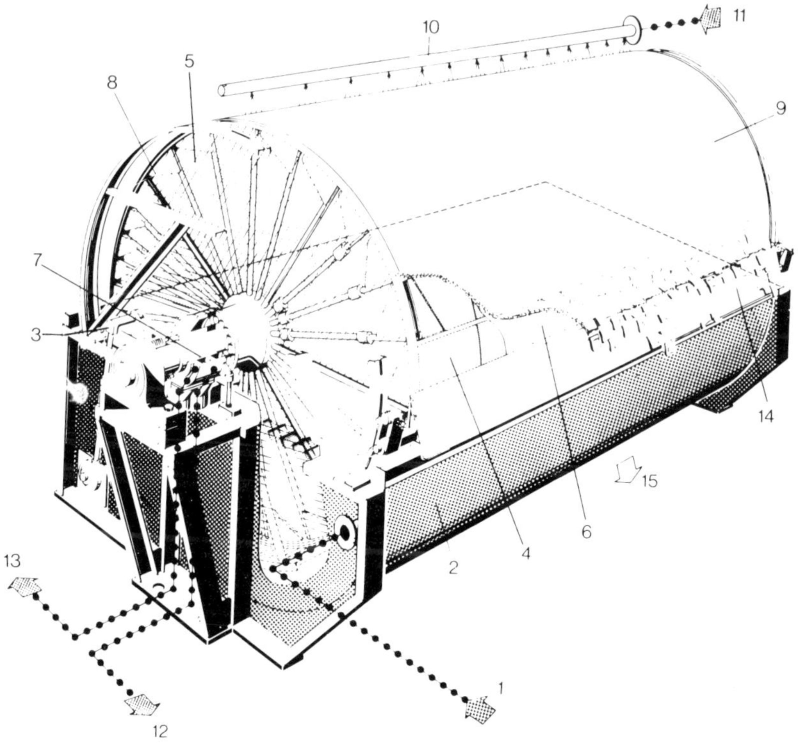

1 Suspension
2 Filter trough
3 Pendulum agitator
4 Filter cells
5 Drum
6 Filter cloth
7 Control head
8 Filtrate pipes
9 Filter cake
10 Washing device
11 Wash liquid
12 Mother filtrate
13 Wash filtrate
14 Discharge device (e.g. scraper discharge)
15 Solid

Figure 6.12 – Vacuum drum filter (Krauss Maffei)

A more recent development performing a similar duty is the vacuum band filter, which has been favourably evaluated for dyestuff isolation in comparison with both the rotary vacuum filter and various filter presses [6]. In filters operating on the moving tray and continuous fabric principle, the vacuum tray is evacuated and filtration takes place while the tray and the fabric are moved forward at the same speed. At the end of the vacuum stroke the vacuum is released and the tray aerated. The fabric is moved forward while the tray quickly returns to its starting position for a new cycle.

6.4.6 Centrifuges

Centrifuge filters are of two main types according to how the solids are removed. In the fixed-bed type the solids cake remains on the wall of the bowl until removed manually or automatically by means of a knife mechanism. In the moving-bed type, more suitable for continuous operation, solids are moved along the bowl using the bowl angle, a scroll or a ram. This leads to a multitude of types, some of which are employed in dedicated intermediates manufacture although less frequently than presses, and vacuum and pressure filters.

6.4.7 Screening filters

The term 'screening' is reserved for filtrations involving the removal of small quantities of catalyst, unwanted insoluble matter or the separation of an adsorbent such as carbon when this is used for the removal of solubilised impurities. Many examples of the previously described product isolation equipment (e.g. presses or centrifuges) may be used as screening filters, but there are also a number of specifically designed items of equipment such as cartridge filters and leaf filters.

6.5 DRYING

Drying is eliminated whenever possible because of energy costs coupled with some inevitable handling losses. There are, however, many cases where products, usually in the form of pastes discharged from a filter, must be dried. Even with optimisation of physical form the water content of pastes varies from product to product in the range 20 to 80%.

The main criteria governing choice of dryer are:

(a) Scale – in terms of both quantity of product and water to be removed
(b) Physical form and handling characteristics of feedstock
(c) Any hazards, or other decomposition problems, associated with product (or vapour)
(d) Physical properties of product
(e) Physical form requirements for subsequent processing or marketing
(f) Costs – capital and operating costs per unit of production.

The following types of dryer are available:
(a) Ovens
(b) Agitated vacuum dryers
(c) Filter dryers
(d) Spray dryers
(e) Drum dryers
(f) Through-circulation dryers
(g) Fluid-bed dryers.

The practical features of these are considered below, leaving the theoretical aspects of drying to more specialised books [4, 7].

6.5.1 Ovens

Ovens of various types are still in general use because of their versatility. They can be used for a wide range of products and for small amounts. The major problems with ovens are their labour-intensiveness and maintaining satisfactory working conditions on loading and discharging. The dried product is invariably in the form of lumps that require comminution. They are not therefore suitable for very large quantities, especially if there is any special physical form required.

Apart from the earlier gas-fired or steam-heated stoves (like large kitchen ovens containing trays of product), variations include electrically heated air circulation stoves (Figure 6.13) and steam-heated vacuum ovens. Even with these refinements, drying cycle times are typically 12 to 36 h.

Microwave oven drying is still at the development stage in terms of dyestuffs manufacture.

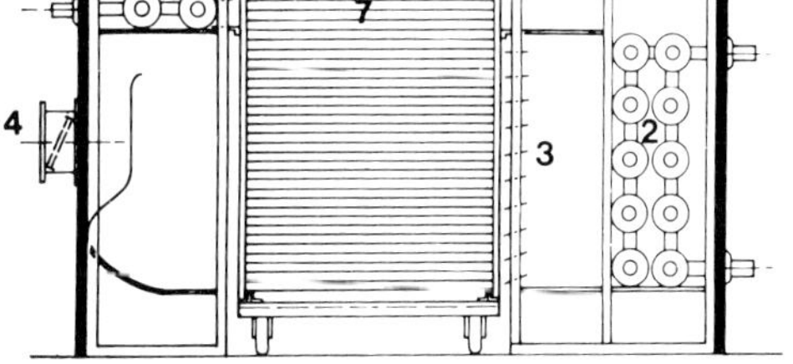

(**1**) High volume air circulation fan
(**2**) Heat source: choice of steam, electric or gas
(**3**) Air distribution louvres
(**4**) Damper controlled fresh air inlet
(**5**) Damper controlled exhaust air outlet
(**6**) Fully guarded fan drive with pre-packed bearings
(**7**) Drying trucks for carrying trays, or direct loading of individual trays for cabinets
(**8**) Standard foot mounted motor.

Figure 6.13 – Cross-section of air circulation stove (APV Mitchell Dryers Ltd)

6.5.2 Agitated vacuum dryers

The main alternative for batchwise drying is an agitated dryer, either in the form of a vertical pan or more usually a horizontal cylinder, operated at reduced pressure. Vacuum operation facilitates containment (e.g. of solvents), and the product is usually a mixture of powder and small lumps which is usable directly or requires only modest after-grinding. It can be used for liquid or paste feedstock but products which dry via a 'sticky' stage are unsuitable. The rapid drying under a wide variety of well controlled operating conditions is particularly advantageous for products that are air- and heat-sensitive.

The paddle dryer shown diagrammatically in Figure 6.14 typifies one specific design for a cylinder type with controlled internal and external heating under vacuum. Economical running costs are claimed for this type of dryer (normal volume from 300 to 16 000 l) at 1.5 kg steam and 0.1 kWh of electric energy per kg of evaporated water.

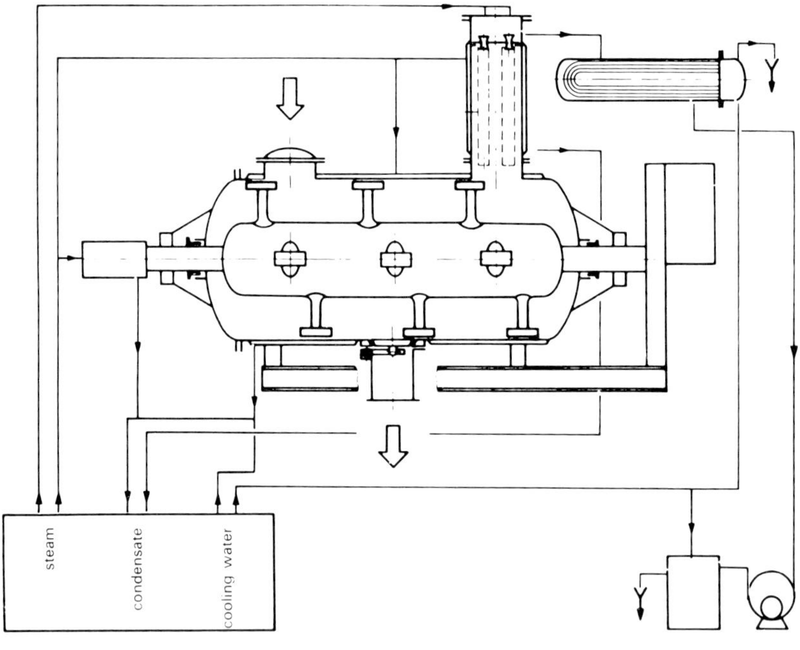

Figure 6.14 – Paddle dryer process diagram (Buss Ltd)

6.5.3 Filter dryers

It is obviously desirable, if possible, to telescope the usually separate batchwise operations of filtration and drying. Filter dryers are comparatively rare, however, since they must meet the criteria for both operations simultaneously. Pressure filters (Section 6.4.2) represent the most notable example of having an optional integrated drying mode.

Best results are obtained when the product is in the form of large particles in a high total solids slurry. This gives the required uniform bed with low resistance to the hot-air flow applied after filtration.

6.5.4 Spray dryers

Spray drying is a continuous operation converting a fluid feed into a dry powder in a single step. This is particularly important for dyes and pigments since precise particle size and low dust level specifications in a safe plant operation under pollution-free conditions can be achieved [8]. It is not the most energy-efficient form of drying, especially for products where the feedstock requires dilution for physical transfer. The compensation for this is the production of a particulate product ready for sale without further grinding. As with all continuous operations, long production runs using optimised conditions offer the best economics.

The flow-sheet shown in Figure 6.15 is a typical concurrent system with atomised feed spray and hot air passing into a chamber where turbulent flow is created. The dried product is collected at the base with any fines entrained in the exhaust gas being captured in a cyclone or baghouse. For

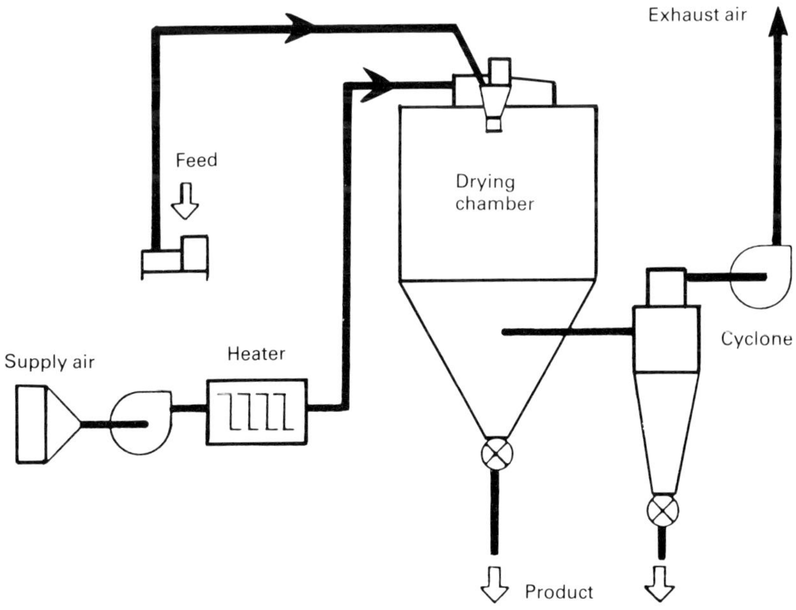

Figure 6.15 – Spray dryer flow-diagram (A/S Niro Atomizer)

energy conservation the exhaust gas may be conditioned in a scrubber/condenser with recycling of the drying gas. Figure 6.16 shows a full-scale tower type nozzle spray dryer with a typical evaporative capacity of about 1 000 kg/h.

The standard design as shown in Figure 6.15 uses a rotary atomiser and produces a dry product with a mean particle size of 50–120 μm. An alternative, producing granules of mean size 200–250 μm, is a tall-form chamber design

Figure 6.16 – Nozzle spray dryer (A/S Niro Atomizer)

using a pressure-nozzle atomiser and a hot-air dispenser system, which creates a streamline flow. In either case the dryer is operated at the highest inlet temperature tolerable (typically 300°C) by each individual product in order to achieve minimum fuel consumption for a given evaporative capacity. Careful monitoring of inlet temperature and flow rate allows the outlet temperature to be controlled (typically at 80°C), which is critical in producing material to a consistent physical specification.

6.5.5 Drum dryers

For drying small or medium amounts of liquids or fine slurries steam-heated drum dryers are moderately efficient but of limited application. They are most suitable for single product continuous operation and are not as versatile as other continuous dryers because of the critical product-specific adjustments that are necessary. Spray dryers are now more frequently installed than drum dryers except in specific cases where, for example, a flake form of product is required.

There are both single- and double-drum dryers, the latter being more versatile since they have the ability to handle pumpable pastes, provided that they do not contain lumps that could damage the drums. The principle of operation depends on the regulated formation of a film on the external surface of a slowly rotating heated cylinder. Figure 6.17 shows a nip-fed configuration as one variant; top-fed, trough-fed or spray-fed options are also available.

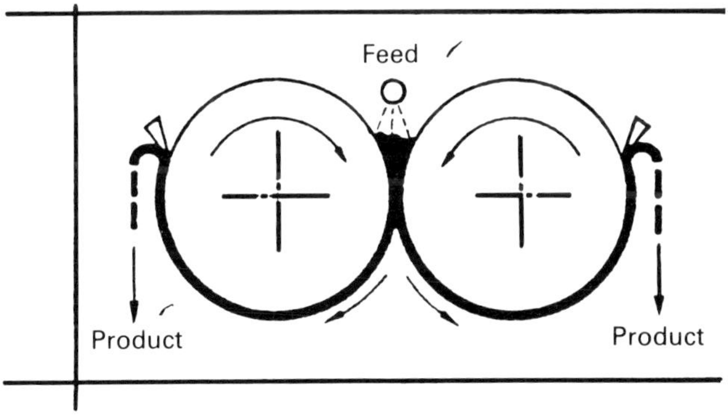

Figure 6.17 – Principle of operation of double-drum dryer

6.5.6 Through-circulation dryers

Continuous tray dryers, in which the material is carried in trays, represent one form of through-circulation dryer, but the term is more usually applied to conveyor-belt types. Particulate material is dried on a moving perforated band usually by blowing hot air through it. Since the material being dried is stationary with respect to the band, the drying operation is analogous to air blowing through a stationary bed of material. For this reason a uniform bed with low resistance to air flow is required, together with a pre-forming device, such as an extruder. The material needs to be in the form of a thick paste, giving a grain structure which is retained at approximately the same size after drying. These specific requirements, coupled with difficulties of containment, severely limit the application of through-circulation dryers to dyes and intermediates.

6.5.7 Fluidised-bed dryers

Fluidising with hot air is an effective way of drying many moist powders and granular products. The advantages are:
(a) Efficient air/solid contact leading to good thermal efficiency
(b) Gentle handling of particles compared with, for example, batch agitated dryers.

Although operation may be on a batchwise basis, fluidised-bed dryers are more effective for large quantities dried in continuous operations. This factor, coupled with the requirement for highly specific fluidisation properties, restricts the use of this type of dryer.

6.6 GRINDING AND MILLING

Particle size reduction is accomplished by a variety of methods that can be categorised as either grinding or milling.

Dry grinding is usually carried out in impact mills (Atritor, KEK or ST); considerable amounts of dust are generated and well established methods are available to control this problem. Dry grinding is an inevitable consequence of oven drying, but more modern methods of drying, especially continuous drying, allow the production of materials that do not require a final comminution stage.

The ball mill, consisting of a rotating or vibrating hollow cylinder partially filled with hard balls, has always been of limited use for dry grinding because of product removal problems. Although historically significant for wet milling, the ball mill has been superseded by sand or bead mills.

Wet milling has become increasingly important for pigments and disperse dyes. Many recently patented designs, particularly from Draiswerke GmbH and Gebrüder Netzsch Maschinfabrik, consist of vertical or horizontal cylinders equipped with high-speed agitators of various configurations with appropriate continuous feed and discharge arrangements. The advantages and disadvantages of vertical and horizontal types have been discussed by Engels [9]; these are so finely balanced as to lead to consideration of tilting versions to combine the advantages of both. Figure 6.18 is a diagram of a

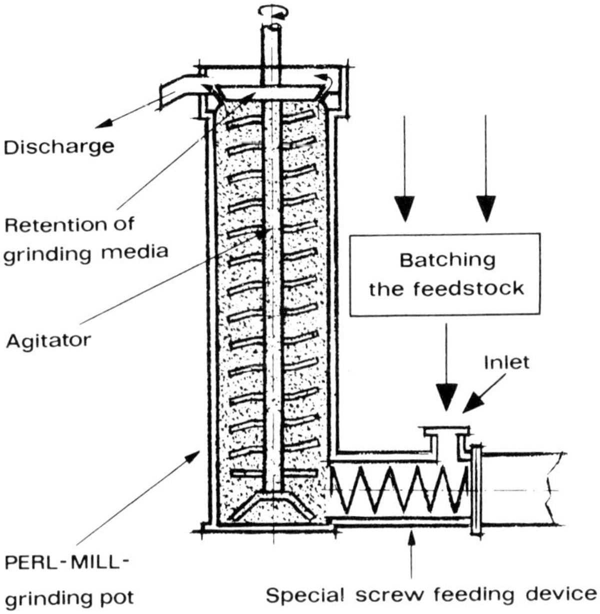

Figure 6.18 – Vertical bead mill (Perl mill) (Draiswerke GmbH)

vertical bead mill showing the agitation arrangement; Figure 6.19 shows a horizontal machine. Both typify the use of small glass beads combined with high-speed mixing. The objective is to achieve the required uniform particle size distribution with the minimum amount of energy input. Various systems are used in the attainment of this objective including a 'pre-milling' stage by integrating a multi-shear mixer into the feed vessel.

6.7 PLANT DESIGN AND OPERATION

6.7.1 Introduction

There are a multitude of factors associated with plant design. Many of these factors, such as:

(a) Economic and technical feasibility
(b) Process materials balance
(c) Energy balance
(d) Kinetics and heat transfer
(e) Instrumentation,

are dealt with generally in books on industrial chemistry [10]. Of the many publications on energy conservation, one review [11] has concentrated on

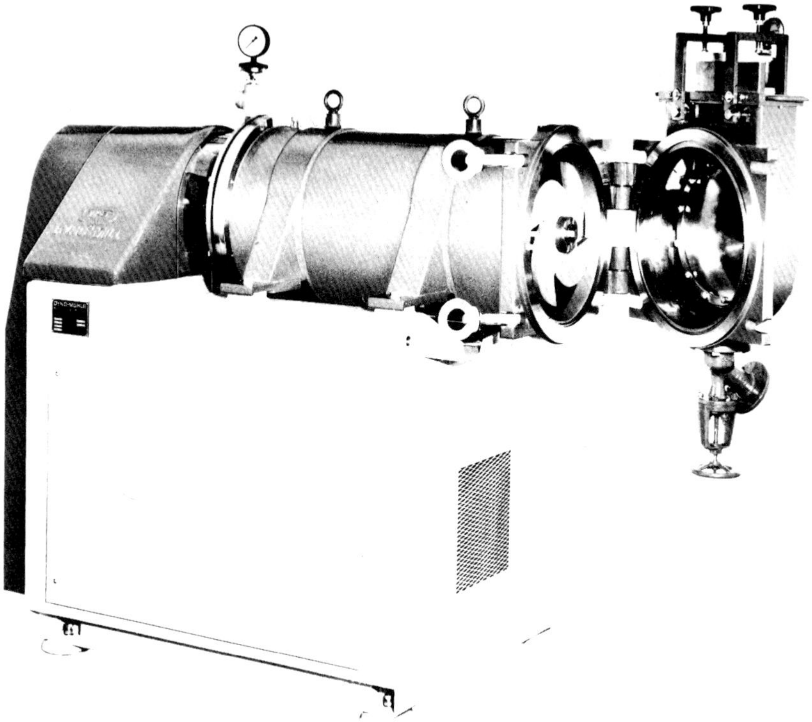

Figure 6.19 – Horizontal bead mill (Dyno mill) (W A Bachofen AG)

batch processing aspects of the dyestuffs industry. The main energy-consuming operations are as follows, each area having been identified as having potential for improvement:

(a) Drying
(b) Milling
(c) Batch processes requiring ice (e.g. diazotisation)
(d) Suspension of solids in the liquid phase.

From what has been written earlier it should be clear that professional process engineers need to be able to combine their academic training with the basic knowledge and experience of the manufacturer, without necessarily being constrained by it. Chemical engineering is a mathematically oriented science, but it should be emphasised that throughout this chapter the approach to the subject has been deliberately restricted to the practical qualitative aspects.

6.7.2 Plant layout

The evolution of plant layouts over the last 50 years has been summarised diagrammatically by Baruffaldi [3], as shown in Figure 6.20. This demonstrates how the concept of total gravity feeding, throughout the sequence charge–react–filter–dry, has gradually been displaced, with increased efficiency, by pumping and other improved methods of mechanised materials transfer. The increase in floor heights directly reflects the increase of the size of installed vessels.

The last part of Figure 6.20, essentially represented in more detail by Figure 5.3, should be regarded as a typical rather than universal arrangement. Other options, including totally horizontal, single-floor layout of the prime operations, are available.

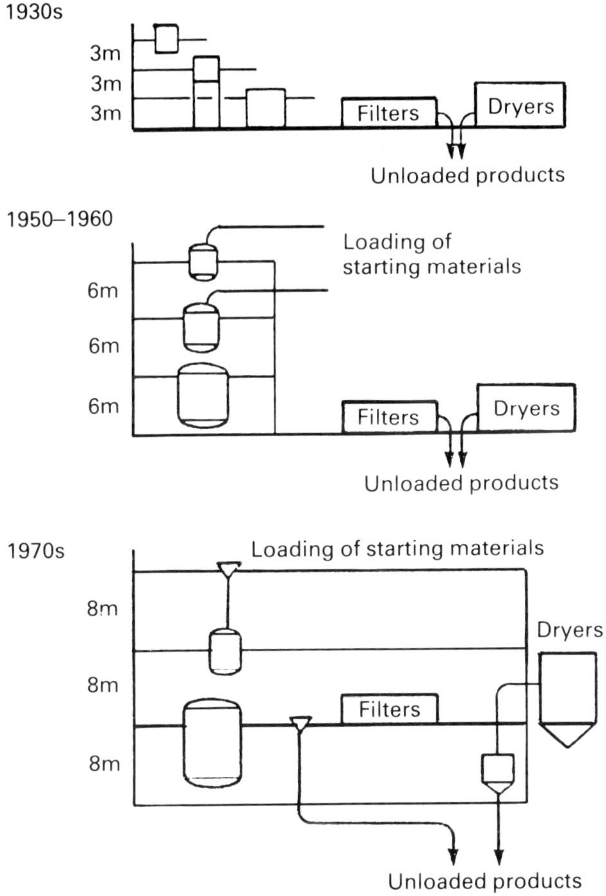

Figure 6.20 – Changing layout of dye manufacturing plants

One design innovation with the latest generation of plants is the building of scale (20:1) models to help in establishing the layout. The significant expenditure on these models is justified by their subsequent use to optimise the disposition of pipework and services with increasing sophistication.

6.7.3 Computerisation

In the past the successful operation of batch processes depended mainly on the skill and accumulated experience of the operator. This operating experience was difficult to codify in a form that enabled full use to be made of it in developing new designs. The gradual evolution of better instrumentation, followed by the installation of sequence control systems, has enabled much more process data to be recorded, permitting maintenance of process variations within the minimum possible limits.

Full computerisation of multi-product batch plants is much more difficult than with single-product continuous units because fundamentally the control parameters vary with respect to time. The first computerised azo [12] and intermediates [13] plants were brought on steam by ICI Organics Division in the early 1970s and have now been followed by many others. The additional cost (approx. 10%) of computerisation has been estimated to give a saving of 30 to 45% in labour costs [3]. However, highly trained process operators and instrument engineers are required.

6.7.4 Plant utilisation

For a single-product continuous plant the availability for manufacture over a given year would be expected to be approximately 95%, the remaining 5% representing approximately two weeks outage for planned maintenance. A single-product batch plant is likely to have a lower availability, around 80%, because of an increased tendency for unplanned maintenance or other forms of breakdown arising from the less cohesive and more interactive mode of operation. With a typical multi-product unit which manufactures, say, ten products, then the availability, and thus the utilisation, may fall to as low as 50%. This figure is mainly due to the loss of production during frequent product change-over, which is a characteristic of this type of operation. Apart from the obvious clean-out and preparation time penalties during product change-over, each new phase of manufacture may initially perform at below full rate until the plant is operating on a routine basis.

The above comparisons are based on a plant operating for a full 168 h per week. Obviously with a five-day working week or alternative modes of operation then the theoretical capacity will be lower, although this could be partly offset by alternative maintenance planning.

In designing a new plant the policy on the mode of operation, together with estimations of the theoretical capacity based on an assumed occupation level, are critical parameters to be considered when deciding the size of a plant. When extrapolated to include consideration of differing equipment needs for a series of unit operations then the calculations required to minimise capital investment for a given production requirement become extremely complex [14].

REFERENCES

1. Brown et al., 'Unit operations' (Wiley, 1950).
2. Booth, Rev. Prog. Coloration, **14** (1984) 114.
3. Baruffaldi, Chim. Ind., **60** (1978) 213.
4. Coulsen and Richardson, 'Chemical engineering, Vol. 2 – unit operation' (Pergamon Press, 1968).
5. Coulsen, Richardson and Sinnott, 'Chemical engineering, Vol. 6 – design' (Pergamon Press, 1983).
6. Sigrist, Chem. Rundschau, **26** (1973) 3.
7. Keey, 'Drying principles and practice' (Pergamon Press, 1972).
8. Masters, Amer. Ink Maker, **54** (12) (1976) 27.
9. Engels, Farbe Lack, **85** (1979) 267.
10. Clausen and Mattson, 'Principles of industrial chemistry' (Wiley, 1978).
11. Kemp and Smith, IChemE symposium series No. 79 (1983).
12. Pinkerton and Robinson, Institute of Electrical Engineers conference publication No. 104 (Oct 1973).
13. Robinson and Trappe, Melliand Textilber., (1975) 557.
14. Flatz, Chem. Eng., (July 1981) 105.

CHAPTER 7

Technical support

7.1 INTRODUCTION

Earlier chapters have made passing reference to several technical activities necessary to support the mainstream manufacturing operation. Further detail is given in this chapter under the headings (a) process development, (b) process hazards, (c) analysis, (d) ecotoxicology and (e) environmental control.

7.2 PROCESS DEVELOPMENT

The major European dye makers spend, on average, about 5% of their turnover on research and development [1]. Those manufacturers that are self-supporting in intermediates will require a higher proportion of process development, compared with the effort they put into new product research, application research and technical marketing support. Where there are equivalent products available from a number of manufacturers, then the process technology component of manufacture is the vital competitive parameter that justifies the continuing search for process improvements.

Apart from the drive for process improvements to reduce manufacturing costs, process development is also required for (a) new products, (b) additional output from existing plants, (c) new capacity and (d) improved environmental and ecotoxicological control. With all these the process developer must have cost benefit in mind. Raw materials costs, even with multi-stage processes, are comparatively easy to derive from yield assessment so that the materials costs for alternative process routes may readily be compared. However, associated processing costs, including the capital allocation, are more difficult. These are based roughly on the hours of occupation required for the process in each plant unit as a proportion of the total time available multiplied by the total cost of operation.

Experiments must be designed, initially at bench-scale, to provide the data upon which to base comparative costings. Uncertainties may require resolution by small-scale plant work on either the process itself or possibly specific associated unit operations, in which direct analogy with other existing products is not possible. A typical scale-down ratio would be 1 l representing a 2 000 l process, with an intermediate simulation step, where considered necessary, of 20 to 40 l.

Scale-up, i.e. duplication of a laboratory process on the plant scale, is a major subject. For the dyestuffs manufacturing industry most scale-up problems arise with either new products or new plants. Initial manufacture of new products, in existing batch units for expediency, requires preliminary development to allow identification of the appropriate plant, and then matching the materials of construction of the plant to the process to avoid corrosion or contamination problems. Physical handling of raw materials and product, including agitation and filtration, must be considered in detail. A new product hardly ever needs a totally new plant; more often one of the unit operational functions of the existing plant has to be adapted or changed.

Major capital expenditure, as distinct from minor modification of existing plant, may be called for when there is a high demand for a particular product, or for other reasons. New plants in contrast to new processes pose many developent problems since simulation on existing plant is often impracticable.

When comparing the potential for continuous or batch operation the following are key questions:

(a) Scale – does the quantity of product (or group of closely related products) justify consideration of a customised continuous unit?
(b) Flexibility – should demand for the product change in future years, is the capacity adaptable compared with that of a batch unit?
(c) Process – given that the reaction rate is fast enough, are the critical parameters of quality (feedstock and product), flow control and mixing capable of measurement and control to avoid any deviation that might lead to 'off-specification' product?
(d) Operation – will the design be able to cope with start-up, shut-down and change-over conditions without significant loss of on-line time?

The reason for the comparative rarity of continuous units in dye and intermediates manufacture is that at least one, and frequently two or three, of the answers to these four questions is in the negative.

Common to all the factors relevant to process development is the need to simultaneously check process safety and product quality; these are covered in Sections 7.4 and 7.5.

A most interesting account has been published [2] covering the development of improved processes for the major naphthalene intermediate 'letter acids'. The development thrust for the new Schelde-Chemie plant was environmental improvement, but the need to optimise capital investment costs, running costs and energy utilisation are also taken into account. In parallel with the process development the unit operations had to be adapted to the state of the art in manufacturing technology. To develop a new plant having 14 000 tonne/year potential capacity for naphthalene intermediates it has been estimated that the effort required is 20 chemists and associated laboratory support for three years.

7.3 PROCESS HAZARDS

All new products and processes require testing for fire and explosion hazards. For processes the thermal stability of reactants, reaction mass and products must be considered in relation to the particular plant occupied. Modified processes must be counted as new ones for safety purposes and the same checks applied. Some of the worst accidents in processing have been due to the failure to realise the impact of an apparently minor change to the process. The possibilities of charging wrong quantities of reactant or the adventitious entry of unwanted chemicals (even water) must all be taken into account.

Hazard evaluation requires systematic programming [3] integrated with the process development work in order to meet predetermined operational standards set by the manufacturer, as well as the legal requirements of the country of operation [4]. Companies with significantly large development programmes covering many products require a hazard evaluation laboratory with special instrumentation, together with appropriate expertise. Measurements and assessments are made to check for process hazards, i.e. potential thermal instability which could lead to runaway exothermic decomposition in the plant, and operational hazards, i.e. assessment of the specific plant to be used, checking, for example, the specific materials of construction for physical or chemical interaction.

Procedures and equipment to meet this need can be developed 'in house' [2] or use can be made of more sophisticated commercially available equipment such as the accelerating rate calorimeter [5] or the Sikarex calorimeter [6].

Nitrations and nitro compounds are potentially hazardous, and there have been many recorded large-scale incidents [7]. Other vulnerable areas include oxidation processes, exothermic processes, processes involving initial total charging of reactants, and processes that rely on vaporisation to provide heat removal. Learning from experience is of paramount importance and recent years have seen a much freer exchange of information between manufacturers, as well as in the number of official incident reports where statutory investigation has been required.

One of the most serious incidents in the industry took place in 1969 in a process involving the diazotisation of 6-chloro-2,4-dinitroaniline. The third batch of a modified process using more concentrated nitrosyl sulphuric acid exploded and resulted in three deaths and 31 people injured. Retrospective testing showed that higher concentrations of diazotized nitroanilines are capable of detonation on heating when the gas cannot escape freely [8]. Although explosions due to detonation are fortunately rare there are many examples of explosions due to runaway reactions and pressure build-up.

The control of vigorous reactions is not the only vulnerable area. When handling powders decomposition may result from applied heat, e.g. from friction, or from sources of electrical discharge. Materials charging and the operations of drying, discharging and packing all need taking into account. A recent fire involving anthraquinone started in a charge chute where part of the metal had been replaced by plastic resulting in a build-up of static electricity, which discharged and ignited the powder. Many products, both intermediates and dyes, have a high available oxygen content (oxygen balance), making them more prone to oxidative decomposition [9]. Screening for decomposition rate, deflagration and, at worst, detonation is essential.

7.4 ANALYSIS

The achievement of higher standards of efficiency in manufacturing processes and higher standards of quality of products in the modern dyestuffs industry would be significantly lower were it not for the many advances that have taken place in analytical and characterisation techniques. Well established spectrographic techniques (e.g. ultra-violet, infra-red and nuclear magnetic resonance) all find application, especially at the process development stage and in analysis for impurities. It is, however, the chromatographic techniques (used for both coloured and colourless products in spite of the name) that are the most versatile since they are also widely adapted for use in process control on a routine basis after the development stage.

The application of all these techniques to the dyestuffs field has recently been reviewed authoritatively elsewhere [10], so only a short summary of chromatographic methods is included here.

Paper chromatography is the oldest technique and simple 'run-outs' on circular paper can still give valuable information. This is especially so for water-soluble dyes where adsorption by the cellulose is an additional factor in the partition process. However, more sophisticated methods with paper strips or thin layers of adsorbent, such as alumina or silica gel, are more commonly used. The latter, named thin-layer chromatography (t.l.c.), is widely used; it gives quick results in the hands of relatively unskilled operators, e.g. in following the progress or measuring the end-point of a reaction on the manufacturing scale. Although other techniques offer higher precision and resolution, t.l.c. is likely to retain its position of importance since it is adequate for so many purposes and has the outstanding merits of simplicity and cheapness.

The use of gas–liquid chromatography (g.l.c.) is restricted to volatile compounds such as amines and phenols. It is of course just as relevant in the detection of volatile impurities in, for example, involatile dyestuffs as in the analysis of wholly volatile compounds and mixtures. Although the use of g.l.c. can be extended to detect volatile derivatives of non-volatile compounds by preparation of volatile derivatives or by pyrolysis, it is usually simpler to use an alternative technique.

The most recent and the most rapidly expanding technique is high-performance liquid chromatography (h.p.l.c.). This is basically optimised column chromatography using a wide variety of adsorbents in finely divided and regular-shaped form. Although requiring more expensive equipment, where applicable it can give rapid and, if required, automated quantitative measurement of component concentrations.

The efficiency of a chromatographic system is measured in terms of theoretical plates (note the analogy with distillation columns) and typical values, demonstrating the impact of h.p.l.c., are:

(a) Simple column 300 plates per metre
(b) T.l.c. 1 500 plates per metre
(c) H.p.l.c. up to 45 000 plates per metre.

The importance of analytical techniques for monitoring the environment for dyes or associated impurities with potentially adverse ecotoxicological effects is mentioned in Section 7.5.

7.5 ECOTOXICOLOGICAL ASPECTS

The Ecological and Toxicological Association of the Dyestuffs Manufacturing Industry (ETAD) was formed in 1974. The 31 member companies represent over 80% of worldwide dyestuffs manufacture (excluding Comecon countries and China) and the two main objectives of the association are:

(a) To coordinate and unify the efforts of manufacturers of synthetic dyestuffs to minimise any possible damage to the environment arising from the use and application of these products and to provide the best possible and practical protection for the users and consumers of these products
(b) To aid and cooperate with government departments and agencies and any other public institution concerned with the ecological and toxicological impact of these products.

These objectives reflect the problems and the increasing pressures that face the dyestuffs manufacturing industry. Historically worker exposure occurred with the two major intermediates benzidine and β-naphthylamine, which are now known to be human carcinogens. Guaranteed safe operating conditions are so prohibitively costly that manufacture has been discontinued in most countries over the last 20 to 30 years. In the UK the use of these two compounds, together with the much lesser used 4-aminodiphenyl, and the manufacture of auramine and magenta are prohibited by the UK Carcinogenic Regulations of 1967, which also control (i.e. registration of manufacture and monitoring of employee exposure) other potentially hazardous intermediates such as α-naphthylamine, *o*-tolidine, dianisidine and dichlorobenzidine.

This situation has led to a somewhat arbitrary distinction between carcinogens and other toxic products, with the UK Health and Safety Executive Guidance Note EH18 on toxic substances using the four groupings:

(a) Proven human carcinogens
(b) Suspect human carcinogens
(c) Substances known to be toxic, but not carcinogenic
(d) Substances not known to be toxic or carcinogenic.

It is necessary to maintain worker exposure as low as is reasonably practicable to all products, even those in the last, and numerically largest, grouping. Ingress to the body is potentially through inhalation, ingestion or skin absorption, and all these should be guarded against. Potential exposure is greater at the manufacturing stage than with the user since the latter has, in practice, much more regulated contact. In spite of this, dye powder contact may give rise to occasional difficulties, and the long-term solution might be to supply dyes in paste or liquid form, ideally in pre-weighed quantities.

Inhalation problems are most severe at the intermediate manufacturing stages owing to the wide range of volatile and potentially hazardous raw materials and primaries. With the most hazardous of these, exposure limits must be laid down and appropriate atmospheric monitoring adopted so that they can be observed.

There remains the problem of correlating laboratory tests and actual human exposure, especially in the assessment of potential carcinogenicity. Exaggerated exposure or injection of chemicals into a wide range of animals may allow potential hazards to be identified, but such results are not necessarily indicative of risk in low-level exposure of humans. Even more tenuous is the link with short-term bioassays that look for cell mutation or damage as a primary screening procedure.

Full epidemiological studies provide the only firm basis for assessing risk to humans. They are comparatively rare because of specificity problems and lack of historical data, and they also suffer from the major disadvantage of being retrospective. They can, however, given a negative outcome, provide valuable reassurance in the event of false positive laboratory results from cell mutagenicity tests. One such study [11] on workers exposed to substituted anthraquinone dyes showed no excess of total or employment-related mortality in spite of a high proportion of the products involved showing a positive Ames (*S typhimurium* reverse mutation) test.

Two major pieces of legislation in this area are the 6th Amendment on dangerous substances in the EEC and the Toxic Substances Control Act (TSCA) in the USA. Both laws distinguish between existing chemicals and 'new chemicals' and require the preparation of an inventory and the notification of the latter. The details on the overall implications, together with test data and their interpretation, have been recently summarised by Anliker and Clarke [12]. The analytical aspects of the industry that are specifically relevant to ecotoxicological assessment of colorants have also been recently reviewed by the Analytical Subcommittee of ETAD [13].

7.6 ENVIRONMENTAL CONTROL

The overall materials balance in dyestuff manufacture is difficult to assess. In general, however, the weight ratio of total starting materials to product is high, with a substantial proportion of inorganic compounds being used. Also waste gases and solid wastes are less of a problem than is liquid effluent, which in addition to inorganics contains organic material as both unrecovered 'valuable' product and unwanted by-product. One estimate [1] for dyestuffs production suggests that 3 tonne of waste is generated per tonne of product, with only 0.4 tonne of waste being organic. Intermediates show a more exaggerated picture. In a typical example, 1 tonne of product formerly required up to 13 tonne of starting materials and gave rise to 68 m^3 of waste liquors containing 11 tonne of inorganics. After process development on the same group of intermediates the ratios were reduced to 7.5 tonne of starting materials and only 13.6 m^3 of waste liquors per tonne of product [2].

Apart from the basic need to control effluents within limits (which vary from country to country) for acid, alkali or colour, there are increasingly stringent requirements in most developed countries for 'toxic' metals (mercury, cadmium, copper, nickel, lead, zinc and chromium), lipophilic aromatic amines and PCBs (polychlorobiphenyl compounds). In addition to pH monitoring, in which the most acidic effluents are often segregated for centralised treatment [14], further characterisation is normally by the parameters BOD (biochemical oxygen demand), COD (chemical oxygen demand) and TOC (total organic carbon). Since dyestuffs are not readily biodegradable they make little contribution to BOD. Although COD and TOC figures are small in the inevitably diluted streams, they are lowered even further where adsorption or chemical aftertreatment is employed to minimise colour, which may be a subjectively unpleasant, if not actually a toxic hazard.

Accurate data on colorants discharged to the environment are not available, but estimates suggest 1–2% loss at the production stage compared with 1–10% loss at the user stage [15]. Manufacturers require little encouragement to minimise these losses since they represent, almost literally, money down the drain. Inland factories require some form of adsorption and/or sewage treatment, possibly in conjunction with the local authority, to avoid visible river pollution. Coastal factories have the option of installing a pipeline and disposing of the effluent in the sea.

The criteria for liquid effluents (and, to a lesser extent in the dyestuffs industry, gaseous and solid effluents) are increasingly important in process development for both new products and new plants. Where potential problems exist the current trend is to maximise recycling within the process and, where necessary, to build in a detoxification or neutralising stage as an integral part of the process. Traditional abiotic processes based on carbon adsorption or the use of cheap inorganic oxidising or reducing agents may in future be supplemented by more sophisticated techniques. Examples that are being developed are hyperfiltration using membranes to retain the higher molecular weight organic colorants [16] and catalytic chemical oxidation using atmospheric oxygen under pressure [2].

REFERENCES

1. Schundehutte, Defazet, **33** (1979) 202.
2. Bretscher, Eigenmann and Plattner, Chimia, **32** (1978) 180.
3. Coates and Riddell, Chem. and Ind., (1981) 84.
4. Vogler, '1982 European conference on evaluation of thermic hazards and prevention of runaway chemical reactions' (London: Oyez Scientific and Technical Studies, 1982).
5. Townsend, Chem. Eng. Progress Vol. 73 (1977).
6. Hub, Inst. Chem. Eng. Symposium 49 (England, 1977).
7. Booth, AIChE Progress Manual Vol. 10 (1976).
8. Bersier, Valpiana and Zubler, Chem. Ing. Tech., **43** (1971) 1311.
9. Gibson and Harper, Inst. Chem. Eng. Symposium 68 (Rugby, 1981).

10. 'The analytical chemistry of synthetic dyes', Ed. Venkataraman (New York: John Wiley, 1977).
11. Gardiner, Walker and Maclean, Brit. J. Ind. Med., **39** (1982) 355.
12. Anliker and Clarke, J.S.D.C., **98** (1982) 42.
13. Bankmann et al., J.S.D.C., **100** (1984) 118.
14. Baruffaldi, Impianti, **13** (1980) 509.
15. Clarke and Anliker in 'Handbook of environmental chemistry', Vol. 3A, Ed. Hutzinger (Berlin: Springer Verlag, 1980) 181.
16. Gaddis, Spencer and Wilson, AIChemE Symp. Series, Vol. 5 (1979).

Glossary

Alizarin	1,2-dihydroxyanthraquinone
Amido-G-acid	2-naphthylamine-6,8-disulphonic acid
Anthrarufin	1,5-dihydroxyanthraquinone
Betamine	2-aminoanthraquinone
β-salt	naphthalene-2-sulphonic acid sodium salt
BON acid	β-oxynaphthoic acid (3-hydroxy-2-naphthoic acid)
Bromamine acid	1-amino-4-bromoanthraquinone-2-sulphonic acid
Bromo pink	1-amino-2-bromo-4-hydroxyanthraquinone
Bronner acid	2-naphthylamine-6-sulphonic acid
C-acid	2-naphthylamine-4,8-disulphonic acid
Caustic soda	Sodium hydroxide
Chicago acid (also 2S-acid)	1,8-dihydroxynaphthalene-2,4-disulphonic acid
Chromotropic acid	1,8-dihydroxynaphthalene-3,6-disulphonic acid
Chrysazin	1,8-dihydroxyanthraquinone
1,6-Cleves acid	1-naphthylamine-6-sulphonic acid
1,7-Cleves acid	1-naphthylamine-7-sulphonic acid
Cresidine	3-amino-4-methoxytoluene
Crocein acid	2-naphthol-8-sulphonic acid
DEA	*N,N*-diethylaniline
Diazo-1,2,4-acid	naphthalene-1-diazo-2-oxide-4-sulphonic acid
DMA	*N,N*-dimethylaniline
DMS	dimethyl sulphate
DNB	1,3-dinitrobenzene
DNCB	2,4-dinitrochlorobenzene
DPA	diphenylamine
ETOX	ethylene oxide
F-acid	2-naphthol-7-sulphonic acid
G-acid	2-naphthol-6,8-disulphonic acid
γ-acid	2-amino-8-naphthol-6-sulphonic acid
H-acid	1-amino-8-naphthol-3,6-disulphonic acid
Hydros	sodium hydrosulphite
J-acid	2-amino-5-naphthol-7-sulphonic acid
K-acid	1-amino-8-naphthol-4,6-disulphonic acid
Koch acid	1-naphthylamine-3,6,8-trisulphonic acid

Laurent's (L-) acid	1-naphthylamine-5-sulphonic acid
MCA	*m*-chloroaniline
MPD	*m*-phenylenediamine (1,3-diaminobenzene)
MTD	*m*-tolylenediamine (2,4-diaminotoluene)
5-NAP	5-nitro-2-aminophenol
4-NAPS	4-nitro-2-aminophenol-6-sulphonic acid
Naphthionic acid	1-naphthylamine-4-sulphonic acid
Naphtol AS	3-hydroxy-2-naphthoic acid anilide
NW (Nevile and Winther's) acid	1-naphthol-4-sulphonic acid
OAP	*o*-aminophenol
OCA	*o*-chloroaniline
OCNB	*o*-chloronitrobenzene
ONA	*o*-nitroaniline
OPD	*o*-phenylenediamine (1,2-diaminobenzene)
ONT	*o*-nitrotoluene
Oxy-Tobias acid	2-naphthol-1-sulphonic acid
PCA	*p*-chloroaniline
PCNB	*p*-chloronitrobenzene
PDCB	*p*-dichlorobenzene
Peri acid	1-naphthylamine-8-sulphonic acid
PMP	1-phenyl-3-methyl-5-pyrazolone
PNA	*p*-nitroaniline
PN salt	4-chloronitrobenzene-3-sulphonic acid
PNT	*p*-nitrotoluene
PPD	*p*-phenylenediamine (1,4-diaminobenzene)
PTSC	*p*-toluene sulphonyl chloride
Quinizarin	1,4-dihydroxyanthraquinone
R-acid	2-naphthol-3,6-disulphonic acid
S-acid	1-amino-8-naphthol-4-sulphonic acid
Schaeffer acid	2-naphthol-6-sulphonic acid
Soda ash	anhydrous sodium carbonate
Tobias acid	2-naphthylamine-1-sulphonic acid